THE LAWN

City and Guilds Leisurecraft Books

SERIES EDITOR : ALAN TITCHMARSH

THE LAWN

George R. Shiels

WORLD'S WORK LTD

Photograph acknowledgements
All photographs are by the author
except those listed below:
Mr A. M. Clarke, Maldon, Essex   33 (left)
Gardena, Royston Road, Baldock, Herts   36 (top right)
Hayters, Bishops Stortford, Herts   40 (bottom)
May & Baker, Romford, Essex   37 (all except bottom right)
Rolawn Turf Producers of York   35
Toro Irrigation, Ringwood, Hants   36 (top left)
Wolf Tools, Ross-on-Wye, Hereford   36 (bottom left), 39, 40 (top left)

Design and illustration
by Victor Shreeve

Published in conjunction with
City and Guilds of London Institute
by World's Work Ltd
The Windmill Press
Kingswood, Tadworth, Surrey

ISBN 0 437 02400 8
Printed in Great Britain by BAS Printers Limited
Over Wallop, Stockbridge, Hampshire

CONTENTS

TABLES OF IMPERIAL/METRIC MEASUREMENT EQUIVALENTS

LENGTH

Imperial	Metric
$\frac{1}{8}$ in	0.3 cm
$\frac{1}{4}$	0.6
$\frac{3}{8}$	1.0
$\frac{1}{2}$	1.3
$\frac{5}{8}$	1.6
$\frac{3}{4}$	1.9
$\frac{7}{8}$	2.2
1	2.5
2	5.0
3	7.5
4	10.0
5	12.5
6	15.0
7	17.5
8	20.0
9	22.5
10	25.0
11	27.5
1 ft	30.0
$1\frac{1}{2}$	45.0
2	60.0
$2\frac{1}{2}$	75.0
3	90.0
4	1.2 m
5	1.5
10	3.0
15	4.5
20	6.0
25	7.5
30	9.0

WEIGHT

Imperial	Metric
$\frac{1}{2}$ oz	14.2 g
1	28.4
$1\frac{1}{2}$	42.5
2	56.7
$2\frac{1}{2}$	70.9
3	85.1
$3\frac{1}{2}$	99.3
4	113.4
6	170.1
8	226.8
10	283.5
12	340.8
14	397.6
1 lb	0.45 kg
2	0.91
3	1.36
4	1.81
5	2.27
6	2.72
7	3.18
8	3.63
9	4.08
10	4.54

VOLUME

Imperial	Metric
$\frac{1}{2}$ pt	0.3 litre
1	0.6
$1\frac{1}{2}$	0.9
2	1.2
$2\frac{1}{2}$	1.5
3	1.8
$3\frac{1}{2}$	2.1
4	2.4
5	3.0
6	3.6
7	4.2
8	4.8
9	5.4
10	6.0

Before you get too involved in the do's and don'ts of lawn care and construction you should understand the role or function of a lawn within the design of a garden.

Does your garden already have a lawn? If it does, what value has it and what is it used for?

If you are currently considering whether or not to have a lawn you might find it interesting to look at the following reasons why it might be a good idea to construct one.

Why do you want a lawn?
Is it because
a everyone else in the area has one;
b you already have a lawn-mower and wish to put it to use;
c you want somewhere for the family to relax or play outdoors?
Should your answer not be one of the above three I hope that the reason you have, or wish to have, a lawn is that you firmly believe it will add an important feature to the garden. The lawn can create harmony between the other garden features; it can create a feeling of space in even the smallest garden; and, provided it is well cared for, the lawn can be a beautiful feature in its own right.

If you are to produce the type of lawn to meet your needs then you must understand why you want a lawn in your garden. The reasons why will determine the type of lawn you build and how to maintain it. When I first had to make the decision about having a new lawn in the garden my answers were that the lawn had to be a play area and also that it had to provide access to the other areas of the garden where choice plants were being grown. Grass in this case had to be green and short to satisfy my needs. Your case might be far more complex and so I'll discuss the types of lawn you might create.

Your garden might be the size of a postage stamp or it might be half an acre – the size doesn't matter at this stage but it will determine how much time you will need to devote to maintaining the grass.

THE IDEAL LAWN
Nearly every garden has a lawn of some description. The ideal lawn is one which suits your needs exactly. It can be a lawn for the children to play cricket on but it can

still be attractive to look at; it is not impossible to have a lawn that is both functional and ornamental.

The presence of dead, yellow or irregular patches of grass and weeds will detract from the overall quality of the lawn and should be avoided by careful maintenance and sound construction. There are always exceptions to this rule and there are many lawn areas full of bare patches and weeds which still serve a purpose. These areas are grassed but can hardly be called lawns.

Always aim at the best you can afford. The type of lawn which you eventually produce must suit your needs. It must also
a suit your site and soil;
b not cost more to produce or maintain than you wish to spend;
c be possible to maintain correctly within the time you have available.

TYPES OF LAWN
There are three basic types or grades of lawn:
1 the very fine lawn;
2 the average or utility lawn;
3 the hard-wearing or second-rate lawn.
Each grade of lawn will serve a function within the design or plan of your garden. The important consideration is whether the lawn meets your needs.

Go back to your answers to the questions about the value of your lawn or why you want a lawn. Remind yourself about the functions of a lawn in your garden, and then read the remainder of this chapter and consider which type of lawn best suits your garden.

1 The very fine lawn
There can be no more beautiful site in the eyes of a gardener than a colourful garden in harmony with a rich green, fine-textured lawn which is criss-crossed by light and dark green mowing stripes. This type of lawn is the ultimate in terms of quality but it is also the ultimate in terms of the thought and effort which are needed to create and maintain it.

The very fine lawn is the classical one which is admired by so many gardeners when they visit well-kept stately homes. The surface is rich green and the grass is very fine and almost velvety in feel. Fine lawns are composed of fine grasses, usually those known as bents and fescues. You can find out more about these in Chapter 5.

If you're a keen bowls player or golfer you could use the fine lawn as a practice green and get extra benefit from your hard work.

Now before you get excited and fired with enthusiasm to go out and build yourself a bowling green you must realise that this lawn, though beautiful to look at, is not hard-wearing. The fine lawn is not the ideal playground for children. It requires careful attention and controlled use if it is to survive. You must not neglect its maintenance or it will deteriorate quickly.

Would a very fine lawn suit your garden?

Remember that this type of lawn is meant as an ornamental feature rather than an area to walk or play on.

In a large garden the fine lawn may be ideal but in the average small garden you would be wise to avoid this lawn altogether. Small gardens often have restricted corners or edges which can be reached only by crossing the lawn. Do this often enough over the same spot of grass and you will create a path through the lawn. Instead, go for something which is more practical and hard-wearing.

If all you require is an ornamental feature then go ahead and plan to make the finest of lawns (see Chapter 2, 'Planning the lawn', and Chapter 5, 'Which grass?'). If you have the space you might even be able to create vistas or you might use the lawn as a backcloth for specimen flowering shrubs.

2 The average or utility lawn

The average lawn is composed of the same types of grass which made up the very fine lawn but it also includes others which are very different.

If the lawn you want has to be velvety smooth – a feature in its own right – then this is definitely not the one to choose but it is infinitely more hard-wearing. In addition to fine grasses it will contain coarser types which are very strong-growing but which do not lend themselves to close mowing. These grasses are used because of their tolerance of wear (see Chapter 5, 'Which grass?').

Although this lawn lacks the velvety texture of the very fine lawn, it can still look attractive. Children's games, bicycles, tents and even the washing line can all be tolerated by this sturdy sward. Its virtue is that it can take most things in its stride; it can even tolerate some neglect.

The average lawn does not need such high levels of maintenance as the finer lawn but it does need to be cared for correctly if it is to give its best. When the weather is warm and moist the lawn will grow rapidly and must be cut regularly to maintain its appearance. Watering and feeding will improve the quality of this lawn and the use of a cylinder mower will create the striped finish which we associate with finer lawns (see Chapter 7, 'Looking after established lawns').

3 The hard-wearing or second-rate lawn

Look carefully at your neighbour's garden or your own if you already have a lawn. Can you see any of the following?

a weeds;
b brown or bare patches;
c thin areas;
d pale green areas;
e pools of water which stand for a long time after rain;
f birds, especially starlings, pecking feverishly at the turf;
g broken, overgrown or uneven edges to the lawn.

If you can see any of the above features you are looking at a second-rate lawn, the most common type in this country.

Second-rate lawns are the product of neglect. They can often be improved but occasionally have to be rebuilt. If rebuilding is necessary there will be a great deal of hard work involved over a relatively short period of time. Have you the time, the will and the energy needed? Chapter 4, 'Preparing the lawn site', will give you an idea of what rebuilding a lawn involves.

If the lawn can be improved by correct maintenance there will still be a lot of hard work involved but it can be spread over a longer period of time. However, the improved levels of maintenance must be sustained, otherwise the lawn will deteriorate again. (See the section on the neglected lawn in Chapter 7, and Chapter 8, 'Lawn problems'.) Are you prepared to sustain the effort? If not, don't start.

WHAT SHOULD YOU DO NOW?

Hopefully you will now appreciate some of the reasons for having a lawn in your garden and you should be aware of the different grades of lawn which can be made. Construction can be exactly the same for each grade of lawn but invariably the very fine lawn is the one which receives most thought and planning.

Anyone can achieve very bad results with little or no thought, but to achieve something which is worthwhile and of lasting value requires careful planning and definite targets to aim at.

Planning is vital to success yet we so often ignore it and simply leap into the work totally unprepared for what we face. Before you get involved in any work at all, read through Chapters 2, 4 and 7 which deal with lawn planning, construction and maintenance. After digesting these chapters you should be in a good position to make a sensible decision about the work you can hope to achieve and thus the type of lawn which you can maintain.

You cannot spend the next few weeks just reading about the work. You have to get down to it. What you can do now will depend on the time of year, the weather and what it is you want to achieve.

If you already have a lawn – do you need to improve it?
– do you need to rebuild it?
If you intend making a new lawn – is it already planned?
– is it to be planned?

Building and rebuilding involves a lot of hard work but whatever the time of year there is still something you can do to make progress. Look at Figure 1.1, which provides an easy-to-use guide to the work you can do now. For further guidance look back to the Contents.

Organise your work carefully, divide the work into logical steps and make sure that you have enough materials to complete each stage. Make sure, too, that you have the right tools for the job before you begin (see Chapter 3, 'Tools and equipment').

CHECKPOINT

Try to answer the following questions without referring to the chapter.

1 What value has a lawn in the design of a garden?
2 What are the tell-tale symptoms of a second-rate lawn?
3 What are the main differences between the finest ornamental lawn and the average or utility lawn?
4 At which times of the year should construction work proceed and when should you devote your time to planning?

Now look back to check your answers.

Figure 1.1 Are you considering building a new lawn? Would you like to improve an existing lawn? If you answered 'yes' to either question, look at the diagrams here to find out what you can do.

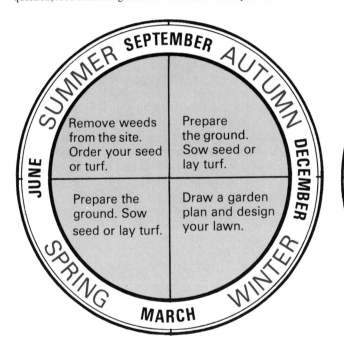

For a new lawn or rebuilding a new one

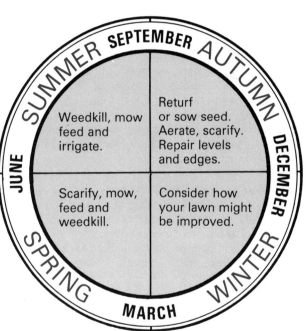

Looking after an existing lawn

CHAPTER TWO

Planning the lawn

Eight steps in planning/Identifying potential problems
Seed or turf?/Checkpoint

I have already stressed the need for careful planning if you wish to create something of lasting value. You must be sure that the lawn will fit comfortably into your garden. Investigate all potential problem areas now to minimise any problems in the future. You should begin by getting to know the site, your garden.

EIGHT STEPS IN PLANNING

Step 1: Scale plan

As the first step in planning – the most vital part of lawn construction – go out into your garden and examine it carefully. Make sketches and notes, and take measurements so that you can
a draw a scale plan of the garden on graph paper;
b mark all existing features – paths, trees, etc. – on the plan.
Figure 2.1 shows a sample scale plan.

As soon as your plan is complete study it carefully. Does a lawn really fit into the design of the garden? Would a lawn add to the beauty of the garden? The vast majority of us will say 'yes' to both questions and so we can proceed to the next and more detailed stages of the planning process.

Step 2: Shape

Look at the plan of your garden again. Are there some features in the garden which are very important in the existing or proposed layout? These important features may determine the shape of your lawn.

Many gardens are square or rectangular, their shape being fixed perhaps by a fence or hedge. For simplicity you could make a square or rectangular lawn. It will certainly fit in with the rest of the garden and will be very easy to maintain. The trouble is that straight lines and rectangles are characteristic of formal gardens and most plots have an informal layout. Square or rectangular garden lawns are not the most attractive shapes to choose.

A lawn which consist of gently flowing curves is much more versatile than a formal rectangular or square shape. The curves can be used in the overall garden design to emphasise other garden features such as groups of shrubs, a pond or a rock garden (see Figure 2.2).

The size and shape of the lawn in a small garden may be rigidly fixed by the size and shape of the garden itself. On modern housing estates the builders seem to leave a small, square piece of garden more by chance than design. Even this tiny plot can be improved by the inclusion of a lawn but a great deal of thought is needed to do the job well.

Your final design will reflect your personality and thus

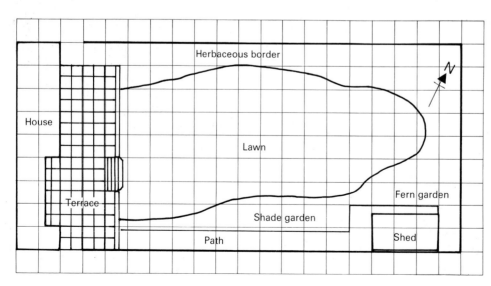

Figure 2.1 Draw a scale plan on squared paper to show all proposed and existing features.

your garden will be unique. The design might embody features from other gardens but how they are used and fitted together will be your own doing. Test your ideas on others before finally deciding on the shape. You'll have to live with your design for some years, so make it the best possible.

Step 3: Light and shade
To be at its best a lawn must have full exposure to daylight. Sunshine for just a few hours each day is enough for healthy growth, but if the site or part of it is shaded for most of the day then you may have potential problems. Rather than growing grass in dense shade, consider the use of shrubs or even herbaceous plants, as they will do far better.

If trees are to form part of the landscape in your garden – and they often do in even the smallest of gardens – remember that the tree canopy could eventually pose problems for the grass immediately beneath it. Do not extend the lawn too far under the shade of trees. Grass can be made to grow in shady conditions but it is hard work and not often worth the effort.

Grasses that are forced to grow in very shady conditions are at a distinct disadvantage because they do not receive enough light to be able to produce sufficient food for growth. The only way to solve this problem is to let the grass grow longer than normal: in this way there will be a greater leaf area for food manufacture and so the plants will have a greater chance of survival.

If the shade is caused by trees overshadowing the lawn there will be other potential problems which must also be considered. Tree roots will compete for the available food and water in the soil, and they will also grow on the surface of the lawn where they will be skinned by the mower blade. The grass which is closest to the trees may suffer excessively and will quickly turn brown in hot or dry weather while the rest of the lawn is still green. Roots eventually die and as they do they will begin to rot, fungi will be attracted to the rotting roots and you may find that your lawn suffers an invasion of mushroom-like fungi. Fairy rings could well develop and disfigure the lawn.

Under very dense canopies you should consider growing shade-loving plants rather than grasses. You could, of course, remove the lower branches of the trees. This would let enough light on to the grass to keep it healthy and dense rather than thin and yellow.

On your sketch plan indicate the position of north. Remembering that the sun rises in the east and travels via south to the west where it sets, do any parts of your garden seem likely to be continually shaded from sunshine? Sketch the deeply shaded areas of your garden on to your plan and make any necessary adjustments to the shape of your lawn.

Step 4: Access
Access will always be required, even if it is simply to allow you to take the mower on to the grass.

Aim to avoid creating very narrow or restricted points

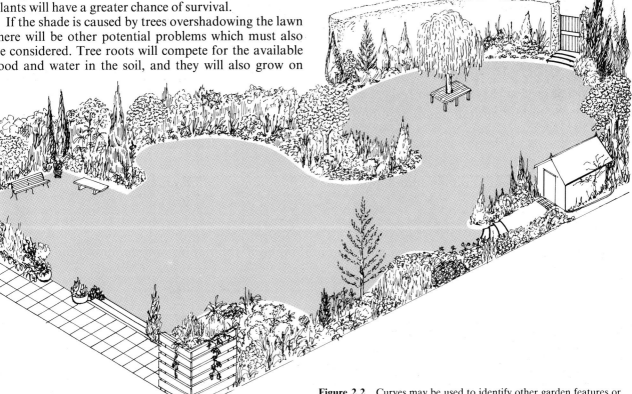

Figure 2.2 Curves may be used to identify other garden features or to create an atmosphere of informality.

of access on to the lawn because this will tend to concentrate wear into a very small section, creating bare patches and making renovation or repair work more difficult.

To provide the best possible access, aim to have at least one open side of lawn which allows free access. This will spread out the wear and will also have the effect of making the lawn seem a little larger than it really is.

Step 5: Paths
Paths are essential in gardens for ease of movement and access. They should not, however, lead directly up to the edge of a lawn and then come to an end. Where a path meets a lawn there will inevitably be an area of severe wear.

For similar reasons avoid siting paths so that they encourage short cuts across the grass. Where a path has to be made so that it follows a severe curve or a very acute angle, a carefully positioned shrub may be all that is necessary to keep all traffic to the path.

There are occasions where a path needs to cut right through the lawn. Stones or slabs can be laid directly into the grass so that they are slightly below the level of the lawn. Position the slabs to allow easy mowing and to ensure that you don't have to stretch too far to reach each slab.

Bearing these points in mind, make any alterations to the paths on your garden plan.

Step 6: Other features
The next step in your consideration of what your lawn should be like must take into account other features, for example whether you wish to include flower beds or even shrubberies in the lawn.

If your lawn is very large it may be possible to include beds of flowers in the design and actually improve the design of the lawn. Small gardens are not quite so easy to deal with. Flower beds within a small lawn tend to emphasise its smallness and they can also make mowing a very difficult operation. The larger the bed or the more beds you have, the more work you create because every flower bed increases the length of lawn edge which has to be trimmed regularly.

A well-designed lawn can be used to generate a feeling of spaciousness even in the smaller garden. By careful use of curves and perimeter plantings the garden can be made to seem bigger than it really is.

Do you want to include flower beds within your lawn?
Perhaps your lawn is there just to provide all-round access so that you can view the flowers at their best from all sides. If beds are a must keep the number down to one or two.

The size of the beds must be kept in proportion to the rest of the lawn and garden. Never place a bed in the centre of the lawn.

When can the beds be made?
Beds should always be cut out of established grass. You must therefore first make a complete lawn without beds. When the grass has become established you will find it possible to create a neat edge to the bed that will not crumble too quickly.

Sketch the shape of the beds on to your scale plan of the garden. Do the shapes fit? Are the beds in keeping with the rest of the design?

Remember that you have to mow the grass. Design the beds so that a mower can get around them easily. Tight corners and very narrow angles can make mowing very frustrating and often untidy.

Position the beds so that they do not create narrow strips of grass at the edge of the lawn; aim to leave at least 1 m for easier mowing (see Figure 2.3).

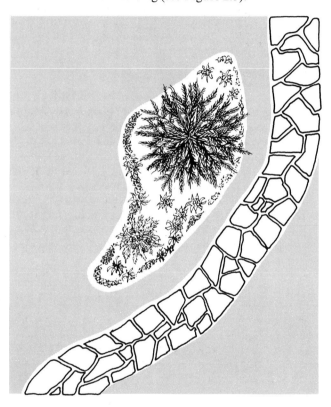

Figure 2.3 Avoid siting flower beds where they create narrow strips of grass which will be difficult to mow.

What should I plant?
Remember that what you decide to plant can have a profound effect on the quality of the lawn. Trees will shade the grass; shrubs or trees can create a lot of work in leaf or blossom clearance. All roots will compete for water and food. Avoid plants which are very vigorous and those which cast dense shade. This reduces your choice a little but will save much frustration and annoyance in the future.

Step 7: Awkward-shaped gardens

Not everyone has the 'average' garden: some will undoubtedly have very difficult and awkward shapes. Unusual shapes are often much easier to deal with because they seem to encourage creativity in design. The worst shapes to deal with seem to be those which are long and thin or short and wide.

Long, narrow gardens can often be improved by using a lawn with large sweeping curves. Shrubs placed in the curves can create a feeling of greater width as well as a feeling of expectancy – what lies beyond them in the distance? Choose very showy shrubs that will catch the eye of the visitor rather than allowing his gaze to concentrate on the narrowness of the garden.

Short, wide gardens can best be improved by planting beds on three sides of the lawn which forms the central core of the garden. Arrange the plants so that the smallest is at the front and the tallest at the back. This will create a sense of depth to the whole garden and even the lawn.

Step 8: Trimming

Will there be a lot of hand trimming to do?

All lawn edges need to be trimmed regularly if they are not to become messy and untidy. Where gardeners forget to make clean edges, they invariably make more work for themselves.

If a lawn is allowed to grow right up to a wall, fence or even a path there will always be long grass at the junction between the two. This long grass will eventually have to be trimmed by shears or a nylon cord trimmer.

Rather than creating untidy edges, always cut a clean edge with a half-moon edging tool. Cutting the edge too often can gradually reduce the size of the lawn and so you should only do it when absolutely necessary. A clean edge to the lawn can be trimmed easily with shears or an edge trimmer.

Hand trimming may also be needed where steep banks are included in the design of the lawn. For ease of mowing, avoid banks which have a slope of more than one in three. If the slope is very steep you will find it difficult to cut the grass, and the top of the slope may also dry out very quickly and turn brown during the summer.

Review stage: How far have you got?

You should now have decided the shape and size of your lawn. If you think about the planning steps and relate them to your original plan you should be able to finalise the design of your lawn.

No book could ever hope to cover every eventuality in the garden because every garden, new or old, is unique in some way. This uniqueness may be a source of problems for you in the future and so the next step in the planning process is for you to take your finalised design back into the garden and try to pick out its faults.

IDENTIFYING POTENTIAL PROBLEMS

Take your sketch plan into the garden. Does your lawn site exhibit any of the following potential problems?

a Very heavy, perhaps poorly draining, soil.
b Water running on to the garden from nearby land.
c Very light soil which is liable to drying out.
d Very chalky soil.
e Very stony soil.
f Very shallow topsoil or no topsoil at all.
g Very uneven site.

I will discuss these briefly below.

Testing your soil

To find out what kind of soil you have in your garden, take a trowel and collect a few samples from the site of your lawn. Wet the soil and rub it between your finger and thumb. How does it feel?

This test will tell you whether your soil is made up of large or small particles.
– If the wetted soil sample feels sticky you have a clay soil.
– If it feels coarse and gritty you have a sandy soil.
– If it feels soapy you have a silty soil.

Clay soils are composed of the finest of all particles. Such soils may not drain very well at all, they may be difficult to cultivate, and they often need to be improved by adding sand.

Sandy soils usually drain freely but may need organic matter, such as peat, to make them suitable for lawn making.

Silty soils are made up of particles which are smaller than sand but larger than clay. Like clays they can be difficult to drain and cultivate but they rarely suffer severely during drought.

There are many variations on the three basic soil types but it is unnecessary to go into details of these here.

Why test the soil?

If your soil is a heavy, sticky clay you will find that it drains slowly and that it is very difficult to work if wet. When this type of soil is wet keep off it altogether. During the summer the same soil will become very hard and dry and may even develop large cracks due to its ability to shrink and expand according to the weather.

Gritty, sandy soils may suffer few, if any, drainage problems. The problems associated with them are likely to have the opposite effect – drying out. Sandy, chalky and stony soils are not able to hold lots of water and so the plants growing in them need to be carefully managed to prevent excessive drying out. Adding peat or other organic matter will enable this type of soil to hold more water and food. Seed which is sown on to a sandy soil should be watched very carefully and kept moist to hasten germination and establishment.

Silty soils are not as strong as clays nor are they liable to crack quite as severely, but they can pose serious drainage problems if subjected to heavy use when wet. A good general rule to follow with this soil is to keep off it whenever it is very moist or wet.

Drainage

How can I tell if my soil will suffer from poor drainage?
Often you can't. Sometimes you can predict problems but no system is foolproof. If you have just taken over a new garden you may not be able to determine problems quite so easily. If you have an established garden you should already be aware of any very wet spots which might exist. The best way that I know to assess garden drainage problems is to dig some test holes and then examine the soil carefully.

Dig some holes in the lawn site about 60 cm deep. Does the soil exhibit grey or blue colours? Does the soil smell unhealthy? Does the hole fill with water?

If you answered 'yes' to any one of these questions your soil will probably suffer from poor drainage to a greater or lesser extent. You might also look for weeds or plants which indicate a wet soil. Rushes are the obvious indicator plants but buttercups and mosses are more common indicators of wet garden soils.

Topsoil
While you have some holes in your plot check the depth of the topsoil.

The topsoil is the darker layer which forms the upper layer of the soil. If yours is a new garden you may find that there is no topsoil at all, or you may find that it has been covered by subsoil or some form of builders' rubbish.

How level is the site?
Finally, as far as potential problems are concerned, you should decide whether your lawn site is level or whether it ought to be treated in some way to improve the levels. A flat lawn is not necessary but neither is one which resembles a mountain range or ski slope. You are the only person who can say whether the levels are good enough for your purposes so you must check the site carefully.

Some slope will assist drainage but too severe a slope will speed up the loss of water from the site so that little actually gets into your soil.

Each of the problems outlined in this chapter will be dealt with in more detail in Chapter 4, 'Preparing the lawn site'. You must, however, be aware of your own problems so that you can make the best of the advice and guidelines on construction.

SEED OR TURF?
The last decision to make in the planning stage of lawn construction is to decide whether to use seed or turf to establish the lawn. The decision should be made in the early stages of the work because you may need to order your turf supply or your seed. You should prepare a programme of work and decide approximately when you will be ready for the turf or seed. Guidelines to the times taken for various jobs are given in Chapter 6.

If you have already decided to use turf or seed write down the reasons why you made the choice. Compare these with the advantages and disadvantages given in Table 2.1.

Each of the points made in the comparison table is relevant to your choice. I will explain each a little more fully so that your choice may be the best for your garden and your needs.

Cost
Turf obviously costs more than seed because it comes partly grown. Someone has had to care for the grass and thus it costs more to produce than seed. Seed weighing 1 kg will cover about 30 m² of a lawn site. Comparing the costs of this seed with the cost of the best turf available, the seeded lawn may be as much as twenty times cheaper than the turfed lawn. Even the cheapest grade of turf will be five or six times more expensive than seed.

Turf is often disregarded for very large areas because it is so expensive and because of other factors to be discussed later.

Table 2.1 **Seed versus turf**

	Advantages	Disadvantages
Seed	Very cheap. Less work involved. You can choose the type of grass. Seed can be stored easily.	Slow to establish. Weeds may come in when the grass is young. Susceptible to bad weather conditions. Birds may be a problem. Can only be grown at certain times of the year.
Turf	Quick effect. Quick to establish. Easy to create neat edges. No bird problems. Can be laid at any time of the year if there is water available.	Expensive. Some turf is poor quality. Turf may introduce weeds or diseases. Soil type may be poor quality. Skill is needed to lay it well. Cannot be stored for more than a few days.

Speed of establishment
Turf has the distinct advantage that it creates an instant lawn. If you have a new house turf is the quickest way to make the garden more than just a bare and barren plot of builders' debris.

If cared for properly the turf will root through quickly and can be used within a few weeks rather than months, as is the case with a seeded lawn. If speed is important, perhaps for the children's sake, then opt for turf. There are several types of turf and these are discussed in detail in Chapter 5.

Work and effort required
Laying turf can be hard work: someone has to carry each piece and place it in position. This is back-breaking work if the lawn is large.

Pieces of turf can be laid badly with very little thought or effort; to lay them well requires thought and skill. A better effect can be obtained from seed with much less effort. Weeds can be a problem in the new seed-bed and you must not use chemicals on very young grass seedlings because they will probably die along with the weeds. Hand weeding is the order of the day. That, too, can be hard work, but it is still infinitely easier than laying turf.

Bird problems
Birds and grass seed go together like mice and cheese. If you are buying seed make sure that it has been treated with a bird repellant or plan to cover the seed-bed with netting or some other form of protection. Don't sow the seed more heavily than is recommended by the supplier because you will increase the risk of disease. It's much better to resow any patches which fail to grow. Turf has no such problems to cope with but you could be buying turf with built-in weed or disease problems.

Making beds and edges
If you are making the lawn from seed, beds should not be cut out until the grass is well established otherwise the edges will soon collapse. Turfed lawns can be laid to leave space for the beds, although it is best to allow the turves themselves to become established first. The turf can even be laid to form the desired shape of the bed and the edges will be quite strong. If speed is of the essence, go for turf.

Turf quality
Buying turf can be a very risky business. There are some very good suppliers around but there are many very poor ones, offering turves which are full of weeds or which have been grown on poor soils. Consult your local *Yellow Pages* and compare the prices and quality of turves offered by specialist suppliers.

The weather
As a final point, remember the vagaries of the weather. Despite your careful planning all can go wrong at the last minute if the weather turns nasty. Frozen ground or snow can prevent turf-laying, heavy rain can prevent both turf-laying and seed-sowing. If the weather does turn against you your timing can be ruined. Seed can be stored easily until the weather improves but turf is a different matter. If turf is left in stacks it will yellow and deteriorate very quickly. If you have to leave it for more than two or three days the turf should be spread out so that it receives full light and air.

To summarise, when planning a lawn you should consider the following points:
a the shape of the garden and the intended lawn;
b the effects of shade on the lawn;
c access to the lawn and other parts of the garden;
d the positioning of paths;
e the inclusion of beds in the lawn;
f the amount of work involved in maintaining the lawn;
g potential problems with the site;
h whether to use seed or turf to establish the lawn.

CHECKPOINT
Try to answer the following questions without referring to the chapter.
1 What is meant by a formal lawn design?
2 Which shape would best suit your garden?
3 What effects can trees have on the lawn?
4 How can you determine which soil type is present in your garden?
5 Can you summarise the reasons for and against using turf to establish a lawn?

Now look back to check your answers.

CHAPTER THREE

Tools and equipment

Mowers/Watering the lawn/Trimming edges
Brushing the lawn/Rakes/Aeration forks
Feeding/Topdressing/Repairing turf
Applying liquids/Rollers/Checkpoint

To keep your lawn in good condition you must look after it, and to do this correctly you must have the right tools. The purpose of this chapter is to introduce you to the sort of tools that you may need. It should also help you to decide which you really need and which can be safely done without.

As the first step you should make a list of the tools you already have. Some general gardening tools can be used on the lawn. This therefore reduces the need to buy tools that are designed specially for grass.

Look at Chapter 7, which deals with lawn care, to find out the operations that your lawn will require as part of its regular maintenance programme.

List those jobs which should be carried out as part of your gardening year on the lawn, and then list the tools which should be used. Could any other tools do the same work? Your list of lawn-care operations should read something like that given in Table 3.1.

As we look at each in turn I will identify those tools and pieces of equipment which are vital.

Table 3.1 **Lawn-care operations and the tools required for them**

Operation	Tools required
Mowing.	Mower — cylinder or rotary.
Scarifying.	Lawn rake or scarifier.
Aeration.	Aeration fork or garden fork.
Brushing.	Besom or stiff broom.
Watering.	Hose pipe, sprinkler.
Top dressing.	Shovel and brush or lute.
Weedkilling and disease control.	Watering can or sprayer.
Trimming edges.	Edging shears and half-moon iron.
Feeding.	Fertiliser spreader.
Clearing lawn debris such as leaves.	Leaf sweeper or broom.
	Other tools which might be useful: sieve for topdressing, trowel, garden lines or canes, mechanical edge trimmer, lawn slitter, planks for turflaying.

MOWERS

Whatever else you might be able to do without, some type of mower should be regarded as essential. I used to cut my first lawn with shears and would never recommend anyone to follow my example.

You need a mower. You now have to select the type of cut you require, the size of machine and the power source you prefer.

Try to answer the following questions.
a What is the total area of grass to cut?
b What is the maximum distance from your house or power supply to the lawn edge?
c Is there ample storage space for a mower or must it hang on a wall?
d Do you have storage space for petrol?
e Can you cope with engine maintenance for a petrol machine?
f Does it matter if the mower is noisy?
g Do you want a very fine cut?
h Do you intend to cut the grass regularly?
i Do you intend to collect the clippings?
Your answers will assist you in choosing a mower, but first you should read the remainder of this chapter.

Type of cut (see Figure 3.1)
Cylinder mowers. These have a number of narrow spiral blades arranged around a central shaft. The shaft rotates and each blade is arranged so that it passes very close to a fixed bottom blade. The grass is caught between the bottom blade and the rotating blades. Because the blades are spiral in shape they give the mower a scissor type of cutting action. If the blades are sharp and well adjusted this mower will produce a fine cut.

The more blades that there are on the central shaft, the finer will be the cut. Machines with four blades will thus not give such a fine cut as the machine with twelve blades on the shaft. For the finer lawns your mower should have at least eight blades; for the average lawn five or six blades will be adequate.

While they produce a better finish than rotary mowers, cylinder mowers do have disadvantages. If your grass is long a cylinder mower may have difficulty in dealing with it. The blades must be adjusted so that the clearance between the rotating and bottom blades is not too great. A piece of paper inserted between the blades can be a useful guide to checking the cleanness of cut: if you turn the blade over slowly the paper should be cut cleanly at all points along the cylinder. If the paper is not cut cleanly the blades should be adjusted and retested. All adjustments will be explained in your mower handbook and you should always refer to this first. Never carry out maintenance or mower adjustments when the engine is switched on or capable of being started accidentally.

Rotary mowers. If you imagine the blades of a helicopter spinning at high speed you will have a good picture of the action of a rotary mower. The blades should be quite sharp but they rely upon their speed of movement for their cutting action, which is a slashing type of cut.

Rotary mowers may be adjusted by a single lever which raises or lowers the blades, or by four individual levers which raise or lower the wheels. There is no other adjustment to carry out. Rotaries are much simpler to use but the quality of cut is inferior to that of the cylinder mower. If you intend having a super-fine lawn then you must go for a cylinder mower; other lawns can be maintained adequately with the rotary mower.

Some rotary mowers have side wheels as part of their design but you will come across others which have no wheels at all. Instead, they float on a cushion of air rather like a hovercraft and so have become known as hover mowers. Hover-type mowers are very good for rough ground or wet grass areas but the operator must be extra careful to avoid getting his shoes or feet cut by the machine.

Other considerations

If your lawn is smaller than 70–75 m² you should be able to cope with it just using a hand-propelled mower. If the lawn is larger or if you are not as fit as you used to be then choose a powered type.

What source of power is best?

Your choice is between battery, mains electric and petrol. **Battery mowers** are seldom seen today but if you do come across one which has been well cared for then it could be a good buy. Such machines are usually fairly quiet and light and are very easy to use, although there are some machines which have a large battery and can be quite heavy. The battery must be charged and carefully stored when not in use.

Mains electric mowers are extremely popular today and a wide variety can be seen in most gardening shops. Such mowers account for more than two-thirds of mower sales in Britain. The smaller electric models are quiet and easy to use but they are not designed for the larger gardens or for heavy work. Remember that you must have sufficient cable to reach all of the lawn. Mains electric mowers are usually much cheaper than the self-propelled petrol mowers but are less powerful. The cutting width is normally in the 30–35 cm range.

Petrol-driven mowers. Size for size a petrol mower will always be more costly to buy than an electric model but it has the advantage of being free to move in the garden unhindered by a length of cable. Cutting widths are variable from 30 cm to over 1 m. Petrol rotary mowers tend to be very heavy and quite noisy. Some of the larger types have a trailed seat on which the operator rides when cutting large lawns. For even bigger areas there are ride-on mowers which look like mini tractors with a cutting unit.

Figure 3.1 Mowers.

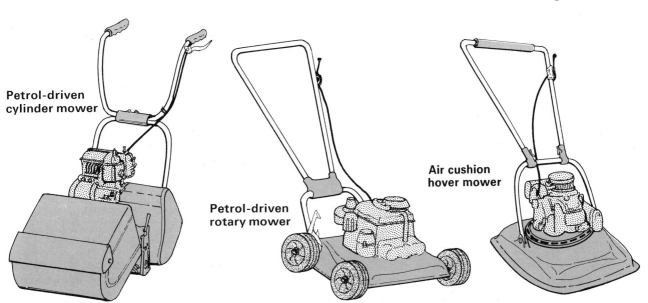

Petrol-driven cylinder mower

Petrol-driven rotary mower

Air cushion hover mower

What size machine do I need?

Size of mower should be related to the size of the garden. Remember, too, that if access to the lawn is restricted at some point this may determine the size of mower. Your mower must be stored somewhere during the winter. The size of the shed or available storage space may also determine the size of mower you should buy.

If the only restriction is how much grass you have then Table 3.2 should be a useful guide.

As a final point remember that narrow strips of grass may also limit the size of mower which you should buy. Other than such restrictions, the larger the mower, the quicker the work will be completed.

What if my lawn is very bumpy?

If this is the case buy a rotary mower, preferably a hover type. A cylinder mower would continually scalp the turf on the bumps.

Are safety considerations important?

All mowers are safe if you follow the manufacturer's guide very carefully. Never adjust a mower with the engine switched on. Make it totally safe by unplugging electric mowers and by disconnecting the spark plug on petrol types. Never let children use the mower.

What about collecting the clippings?

Clippings left on the lawn are unsightly. Many mowers can be bought with a grass box; only the very smallest cannot. A garden rake will quickly collect the clippings on a small lawn. If the clippings are wet you will find that they tend to block the grass box very quickly. Some types of box are always blocked easily and so you should ask to try a mower before buying it.

Can any mower produce a striped finish?

Yes, if it has a rear roller; otherwise only the cylinder mower will make stripes.

Mower maintenance

A few general points should be borne in mind to keep your mower in trim:

Always clean the mower thoroughly after use; a stiff brush is best. Keep battery mowers fully charged and regularly top up the water level in the battery with distilled water.

Check the blades regularly. Blades may become loose or chipped with use. Have defects remedied quickly and never use a mower with a missing or loose blade.

Simple devices can be purchased for sharpening cylinder blades but it is usually better to have a local supplier carry out an annual overhaul on your machine.

The cable for electric models should be checked regularly to ensure that it is not cracked. If you have to make repairs use special waterproof tape.

A petrol mower must have an oil change and it must

Size of lawn (m²)	Recommended size of mower (cm)
500	30
800	35
1,000	40
1,250	45
1,500	60
2,000	70

Table 3.2 **Recommended mower sizes**

have any other lubrication points oiled or greased. All machines should be stored in a dry place after they have been thoroughly cleaned.

During the cutting season you may need to alter the height of cut on the machine. Rotary mowers are easily altered, but cylinder mowers are a little more complex. The process is explained in Figure 3.2.

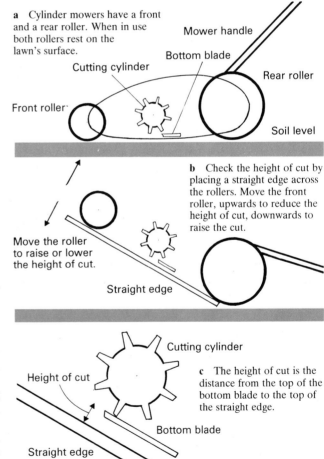

a Cylinder mowers have a front and a rear roller. When in use both rollers rest on the lawn's surface.

b Check the height of cut by placing a straight edge across the rollers. Move the front roller, upwards to reduce the height of cut, downwards to raise the cut.

c The height of cut is the distance from the top of the bottom blade to the top of the straight edge.

Figure 3.2 Altering the height of cut on a cylinder mower.

WATERING THE LAWN

Your choice of equipment will range from the basic watering can right up to automatic sprinklers and self-travelling models suitable for the larger garden (see Table 3.3).

You should always check how much water your system applies. The manufacturer may tell you but I suggest that you examine the soil afer watering for a known period of time to find out for yourself. How deep did the water penetrate? It should be to a depth of at least 15 cm. Look at Chapter 7 for the way to check how deeply water has penetrated into your soil.

Bear in mind the fact that too much water can cause a lot of problems. Incorrect watering is also harmful. Try to minimise the amount of water you use. If yours is to be the very finest of lawns then invest in the oscillating type of sprinkler as it works out as the best buy. If cost is no restriction and you have a very large garden buy the fully automatic system for convenience.

Table 3.3 **Watering systems compared**

System	Advantages	Disadvantages
Watering can.	Cheap with no extra costs.	Time-consuming and impractical for larger lawns.
Hose pipe.	Less laborious than the watering can.	Time-consuming and not very accurate in spread.
Static sprinkler.	Fairly even pattern of spread. Simple and cheap.	Difficult to get the right overlap when the sprinkler is moved.
Perforated or sprinkler hose.	Good for long thin areas.	Your local authority may not allow their use.
Oscillating sprinkler.	Adjustable and fairly accurate. Rectangular pattern of spread which enables accurate overlap.	A licence is needed before you can use it. This applies to other sprinklers too.
Pulse jet sprinkler.	A large area is covered very evenly.	Time-consuming.
Self-travelling sprinkler.	This takes the work out of watering very large areas.	Expensive and not very accurate.
Pop-up sprinklers.	The Rolls-Royce of irrigation systems. A system for the larger garden can be fully automatic. Programmed to switch on and off automatically.	Expensive.

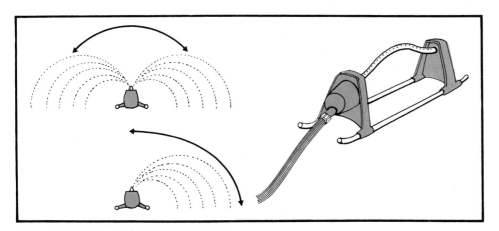

Figure 3.3 Oscillating sprinklers provide fairly even coverage but you should have a licence to use one. The sprinkler slowly turns from side to side to cover a large area but you can set it so that it waters on one side only.

TRIMMING EDGES

The vertical edges of the lawn must be kept well trimmed. Edging shears (see Figure 3.4) may be satisfactory but they can be tiring. If you have a lot of edge to trim a battery- or mains-operated trimmer is the easiest way to do the work. It's not much quicker but it is less tiring. The disadvantage is cost.

Other types of edger consists of a series of sharpened spikes which rotate against a fixed blade. You have to push the tool along the lawn edge and that can be just as tiring as using hand shears.

At least once a year trim the edge with a half-moon edging iron (Figure 3.5). Use a board or string line to make a straight edge to the lawn.

Some edging must be done but you can keep the cost of tools down by using shears. With a lot of care a spade could be used instead of an edging iron to create a smooth edge to the lawn in spring.

BRUSHING THE LAWN

The best tool to use for brushing the lawn is the besom, but if you do not wish to buy an extra tool a good stiff broom will serve just as well (see Figure 3.6).

Mechanical sweepers may also be useful in the larger garden (see Figure 3.7).

RAKES

Lawn rakes (see Figure 3.8) are made with thin wire tines which tear into the grass surface and pull out any weeds and thatch that might be present. A garden rake can be used instead but as the tines are broader it will not be as effective. Specialised tools are available – the scrake for example – but again such tools are useful rather than essential.

AERATION FORKS

Only the purpose-made fork can be used to hollow-tine the lawn if compaction is a problem. This work is only infrequently carried out on the better lawns – never on many lawns – and so the cost of such a fork is rarely justified. Solid tine aeration can be carried out effectively by a garden fork. Shallow spiking can be done quickly with a wheeled machine that has spikes on a rotating cylinder. The machines are usually light and rarely achieve deep penetration into the soil.

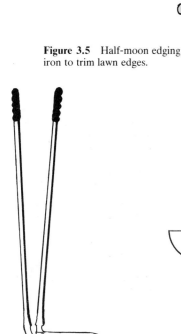

Figure 3.5 Half-moon edging iron to trim lawn edges.

Figure 3.4 Edging shears with handles 1 m long to reduce the need to bend.

Besom

Stiff broom

Figure 3.6 Brushes for the lawn.

Figure 3.7 Mechanical leaf sweeper for the larger garden.

Figure 3.8 Lawn rakes have thin wire tines. Often these rakes are called springbok rakes.

FEEDING

Applying fertiliser can be carried out very accurately by hand but it is time-consuming. Mechanical spreaders are far quicker.

The most common type of machine used for this has a hopper to hold the fertiliser and a notched or ribbed roller in the base of the hopper. As the roller turns, fertiliser collects in the notches, is carried out of the hopper and is then dropped on to the grass evenly. The hopper can be raised or lowered to adjust the rate of fertiliser application. Always check the machine before use on the lawn by spreading fertiliser on to some polythene. Collect up and weigh the fertiliser from 1 m² of polythene. If the rate is not the rate you need adjust the control device and test the machine again.

Spinning-disc spreaders are available for use on larger areas but greater care is needed in their use to obtain even coverage of the lawn.

When using a spreader, switch it off when you turn to prevent some areas being double-dosed. The right and wrong ways to feed a lawn are shown in Figure 3.9.

Hand distribution is still the most common method of applying fertiliser. So long as you are careful, it is very efficient and safe.

TOPDRESSING

Fertiliser spreaders can be used to apply topdressings but, again, hand spreading is just as accurate if done carefully. A shovel and barrow are the only tools you need but you should mark the lawn into strips for accuracy.

A stiff brush is usually good enough for levelling topdressing, but if large areas are involved you might like to buy or make a lute.

The bought version consists of a rectangular metal frame attached to a long metal handle. The frame is pushed back and forth across the topdressing or the soil surface of a seed-bed so that it pushes soil into hollows and thus creates a smooth surface.

You can make a suitable tool by fixing a two-pole handle to a board about 1.5 m long as shown in Figure 3.10.

a RIGHT WAY

Figure 3.10 A lute for levelling topdressing. You can make a lute by fixing a two-pole handle to a 1.5 m board.

Figure 3.9 Feeding the lawn. First treat two strips at each end of the lawn. Then treat the rest of the lawn in strips, allowing a slight overlap on each strip.

Applying fertiliser with a spreader in this manner increases the risk of missing some patches and double-dosing others.

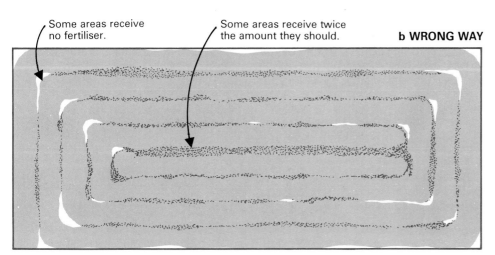

Some areas receive no fertiliser.

Some areas receive twice the amount they should.

b WRONG WAY

REPAIRING TURF

Occasionally you may need to lift turf, to repair edges or worn patches for example. There are special tools for the job but a garden spade is used quite successfully by many gardeners. Turf lifted by hand will have to be trimmed to size and so a box and knife will also be needed (see Figure 3.11). Turf-lifting is only rarely necessary and buying special tools would be wasteful.

Gauge box: 3-sided without a top

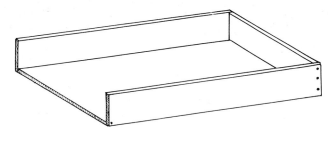

Gauge box with turf in position

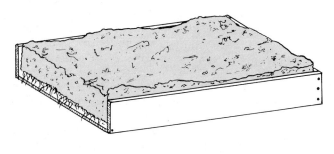

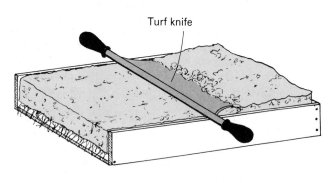

Figure 3.11 Repairing turf. Hand-lifted turf should be trimmed in the gauge box. Place the turf grass-side down in the box and cut off any soil which extends above the sides of the box by drawing the knife across it.

APPLYING LIQUIDS

Weedkillers and fungicides, and even liquid fertilisers, can be applied to the lawn through a dribble bar fixed to a watering can, from a knapsack sprayer or, in the case of weedkillers, from an aerosol spray can.

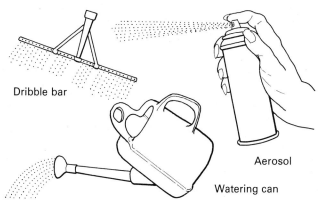

Dribble bar

Aerosol

Watering can

Figure 3.12 Applying liquids to the lawn. Aerosols and watering cans may be used to apply chemicals. A dribble bar may be fitted to the watering can spout to improve the efficiency of spread.

ROLLERS

I will make very little mention of rollers because their use often causes more harm than good. On light soils they do have a place. If you have a cylinder mower – a petrol one in particular – the roller on this machine can often do as much good as a separate roller.

A roller for use on grass areas should not be very heavy. Remember that it is not to be used for levelling the lawn; it simply helps to firm it. My advice if you must have a roller is to borrow your neighbour's. There is a type which has been designed to be filled with sand or water and so is adjustable. This is the best type. The edges of the roller should be rounded to reduce damage to the grass surface.

CHECKPOINT

Try to answer the following questions without referring to the chapter.

1 Which piece of lawn equipment should be used only by those who have a licence?
2 Which type of mower gives the finest cut?
3 What are the disadvantages of using a mains electric mower?
4 What size mower would best suit your garden?
5 Which mower is likely to be the quietest?
6 How can you tell whether or not your soil has been watered adequately?
7 What can be applied by aerosol spray to the lawn?

Now look back to check your answers.

CHAPTER FOUR

Preparing the lawn site

Clearing the site/Grading and levelling the site
Drainage/Cultivating and improving the soil
Final preparations/Checkpoint

By this stage in your reading you should have made several decisions about your lawn. You should have decided:
a whether you really want a lawn;
b the quality of lawn you wish to make;
c which shape and size best suits the rest of the garden;
d the location of the lawn, flower beds and paths which might affect the lawn;
e whether to use seed or turf.
Each and every garden is unique in some way. Each has its own special problems, perhaps its soil or its position. Some gardens are surrounded by hedges or fences; some are situated in hollows so that water drains from every other garden into them.

In Chapters 1 and 2 you studied your plot carefully and hopefully will have identified some of the potential problems. If you follow the stages in construction which are outlined in this chapter you should be able to deal with your problems successfully and will then look forward to many mowing days.

When do I start work?
You will need at least three months to do the work properly and you will need to plan it so that the work can be completed comfortably in the time allowed. If stages of the work have to be rushed there is a risk that they will not be carried out thoroughly. Skimping on the work is likely to result in a poor end-product and more work in the future, so don't do it.

No matter when you decide to work on a new lawn there is always something that you can do, even in winter (see Table 4.1).

Table 4.1 **Preparatory work by season**

SPRING	A good time for sowing seed, preparing seed-beds and cultivating the soil.
SUMMER	The ideal time for fallowing the soil and killing off weeds which might infest the lawn. Complete planning work and even soil cultivations.
AUTUMN	Prepare the soil and sow seed or lay turf.
WINTER	The best thing to do is make plans and keep warm.

What do I need to do to make a lawn?
Making a quality lawn involves careful attention to preparing and improving the soil so that it will grow healthy turf. The operations which are normally required are as follows:
a clearing the site;
b grading and levelling the site;
c draining the site if necessary;
d cultivating and improving the soil;
e carrying out the final work to prepare the seed-bed.

CLEARING THE SITE
Have you recently moved into a new house? Have you recently taken over a neglected garden? Have you decided to build a new lawn or rebuild an existing lawn?

A 'yes' to any or all of the above questions suggests that you have some clearing up to do before you can start building the new lawn. New gardens are often covered with builders' rubble; older gardens may simply be covered with weeds or other unwanted vegetation. Whatever the debris, it must be cleared completely before the next stage of the work can proceed.

Rubble and large stones should be picked up and removed from the site unless they could be of value in building a soak-away for a drainage system. The use of soak-aways is discussed in the section detailing drainage installation. Stones which are left on the surface of the site could eventually come into contact with the blades of your mower and cause serious and expensive damage. If the garden is a new one you may well find whole bricks below the surface of the soil. Very careful checking of the soil is obviously essential if future problems are to be avoided.

Builders often leave heaps of subsoil at intervals over the garden. These too must be completely cleared away. Don't yield to the temptation to spread the subsoil over the remainder of the garden because it will reduce the quality of the topsoil and the future lawn. Unusual situations are often encountered in new gardens. When I took over a new garden I found that the top 10 cm of the soil was subsoil. Below the subsoil there was about 30 cm of rich black topsoil! If you encounter such a situation the only answer is to clear the subsoil away completely – no half measures will do.

Unwanted vegetation, whether grass or shrubs or stinging nettles, should be cleared thoroughly. Burn it if possible and spread the ashes on the site.

Perennial weeds must be totally removed from the soil, making certain that the roots are also dug out. Think twice before deciding to leave trees or shrubs near the lawn: they are an endless source of problems with surface roots and shade causing the lawn to become thin and weak.

If the weed infestation is severe chemical herbicides may be the simplest answer to the problem as removing weeds by hand is very laborious. Much effort can be saved by resorting to chemicals. Provided you follow the instructions given by the chemical manufacturer you should find the chemical safe to use and very effective.

Your choice of herbicide is limited to those which do not persist in the soil for long periods of time. Choose a contact herbicide that kills the tissue with which it comes into contact, usually the leaf tissue of course. Other herbicides which can be of great value are those which are translocated or moved within the plant's system. This type will kill the roots of perennial weeds and can save you much hard work. Herbicides are discussed in more detail in Chapter 8, 'Lawn problems', where the discussion centres on selective herbicides or weedkillers. For ground clearance other types are often used; glyphosate is a very good example. It is a foliar acting herbicide which is translocated. This means that when it is sprayed on to the leaves it is absorbed into the sapstream. It is capable of killing the roots of difficult weeds such as couch grass. Aminotriazole can also be used but several weeks must be allowed to elapse after spraying before any plants can be grown in the soil. For safety, use only those herbicides which do not last in the soil at all.

If you do not like using chemicals at all a flame gun could be used instead but it will kill only the top growth of perennial weeds, not their roots. Do not rely on several passes with a cultivator to kill the weeds: it will do no more than temporarily bury them. The final method which might be employed is to skim the vegetation off with a spade, but again this is very tiring, back-breaking work and will leave many roots undisturbed to grow again.

Figure 4.1 Sand-carpet lawn construction.

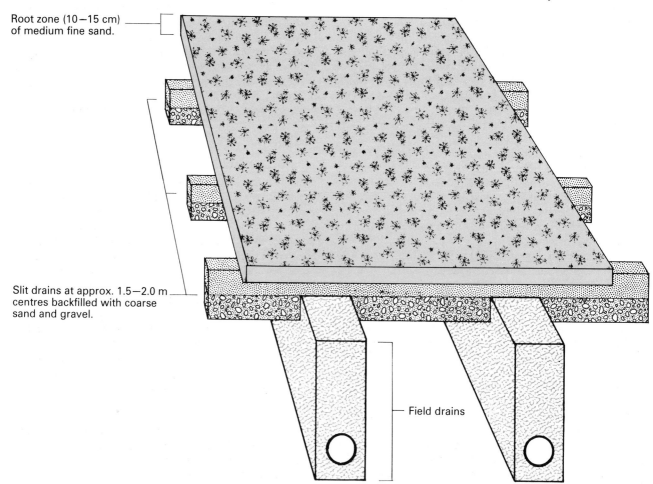

Root zone (10–15 cm) of medium fine sand.

Slit drains at approx. 1.5–2.0 m centres backfilled with coarse sand and gravel.

Field drains

Bare soil

Now that the surface debris and vegetation have been thoroughly removed from the site or burned and spread you should be left with bare soil. This is an ideal opportunity to assess the quality of the surface levels. Are they satisfactory? It is also a good time to check the topsoil depth if you have not already done so. Dig test holes and examine the walls of the hole. How deep is the surface layer of darker soil?

For grass to grow well and not suffer from water or food deficiencies too quickly it needs to have a minimum depth of topsoil of 10 cm but preferably of 15 cm or more. If you find any areas which have 10 cm or less these must be made up before any grass is established on them.

If the garden is fairly level already you may not intend doing any further levelling work. If this is the case you must ensure that every section of the proposed lawn site has enough topsoil. Areas which do not have enough topsoil tend to turn yellow and suffer from drought and even waterlogging far sooner than the rest of the lawn.

What can I do if my garden has no topsoil?
The first and perhaps the most obvious answer is to find a supply of soil and buy enough to spread over the lawn area, ensuring that there is always a depth of at least 10 cm.

But what if no soil is available? Unusual techniques are often the best answer to unusual problems. A technique which I have used to build very fine lawns, where the topsoil was either too thin or non-existent, might be of value to you. We can refer to it as the sand-carpet technique of lawn construction. It is a tried-and-tested method of construction for ornamental lawns and even top-quality golf greens.

Briefly, the sand-carpet system (Figure 4.1) involves the following stages or sequence of work:

a   The existing topsoil or subsoil surface is levelled by moving soil around within the site.
b   Drains are installed and the trenches are back-filled with fine gravel up to 5 cm below the soil surface. The top 5 cm are filled with medium sand. (This is sand with particles of 0.1–0.6 mm in diameter.)
c   A layer of medium sand is then spread over the entire lawn site so that there is at least 10 cm of sand. The sand should be raked level and then a layer of peat should be added and spread over the surface so that there is about 10–12 mm evenly over the sand. Rake the peat into the sand, irrigate and then sow grass seed into the sand/peat mixture.

The sand carpet will be sufficient for grass growth, and it will drain perfectly yet remain firm. Extra care with watering will be necessary but this method can produce a super lawn. The sand which you use is important: it should be a medium-grade sand with most of the particles in the range 0.1–0.6 mm. Methods – seed-sowing especially – are dealt with later in this chapter and in Chapter 6.

GRADING AND LEVELLING THE SITE

Grading and levelling are really two quite distinct processes in lawn construction. Grading is normally taken as referring to the movement of quantities of soil, often large, in order to remove surface irregularities. Levelling is the term which we give to the final stage of preparing a smooth surface. It involves fine adjustments to the surface level using a spirit level and straight edge for accuracy.

Where a site is very uneven both grading and levelling may be necessary in order to create an acceptable surface for the lawn.

Being realistic about the operation of levelling a site, let me begin by stressing that a good lawn does not need to be exactly horizontal or perfectly flat. Contours can be very attractive, especially in adding to the informal look which many gardeners seek, but the contours should be gentle so that your mower will not skim off the top of the slope and leave a bare patch in the new lawn. Smooth, flowing undulations are desirable on some lawns, and some slope can be a definite advantage because it will assist the flow of water across the surface and off the lawn. If banks are to be included in the construction do not create them with a slope that is steeper than 1 in 3 or you will make mowing excessively difficult.

Don't take my comments as an excuse to leave your lawn site looking like part of the Himalayas. Steep slopes and severe hills and hollows cannot be acceptable in any circumstances. Small lawns should be smooth because undulations tend to be out of scale; larger lawns can be contoured without their appearance being spoiled.

Grading

What should I do about large surface irregularities?
Where a site is very uneven it may be necessary to carry out major grading. Major grading can be expensive; it is certainly hard work and very time-consuming. If you examine the site carefully you may find it preferable to create a terraced lawn rather than go to the trouble and effort of grading the site thoroughly. If you do decide to create terraced lawns then remember that you have to maintain the banks and they can be very troublesome.

Major grading work involves a number of operations which are carried out in the following sequence:

a   The topsoil should be stripped from the entire site and then stacked in a convenient spot for future use.
b   The subsoil should now be all that is left. The hollows or humps should be levelled out by moving soil from one spot to another. For greater accuracy use pegs and a spirit level in the way described under the heading 'Levelling' below.
c   Replace the topsoil on the levelled subsoil and ensure that at least 10 cm depth is spread over the entire site. This should be the depth of topsoil after settlement rather than while the soil is still loose and fluffy.

If you do not have anywhere to stack the topsoil you might be able to level half of the site first, stacking the stripped topsoil on the second half of the plot. The subsoil is levelled and the topsoil is replaced. The whole operation is then repeated with the second half of the site. A little more thought and care is required to do the work this way but the results can be just as accurate.

Major grading is hard work and it is very time-consuming but cannot be rushed if the work is to be done properly. See Figure 4.2.

What about minor surface irregularities?

If the uneveness of your site is only slight then the topsoil will not have to be removed and you can breathe a sigh of relief that you have less work to do than might have been the case. Minor grading does not involve topsoil removal on a large scale.

Minor hollows can be corrected by adding soil from another higher part of the site or from a bought-in supply. If you do take soil from high spots to fill low spots do remember that the minimum depth of topsoil is 10 cm but ideally should be 15 cm or more if possible.

If small areas of the site do have less than 10 cm of topsoil you should dig out some subsoil and replace it with topsoil to increase the depth.

If you have the almost perfect site then it may be sufficient to rake out any very small hollows and still produce a level surface.

1 Topsoil earmarked for removal.

2 High spots on the subsoil marked for stripping and moving to low spots.

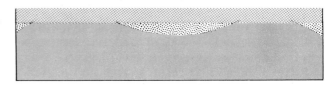

3 Subsoil levelled; topsoil replaced.

Figure 4.2 Major grading stages.

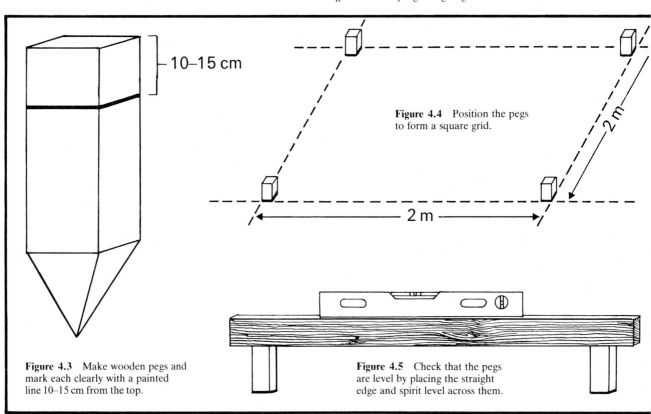

Figure 4.4 Position the pegs to form a square grid.

Figure 4.3 Make wooden pegs and mark each clearly with a painted line 10–15 cm from the top.

Figure 4.5 Check that the pegs are level by placing the straight edge and spirit level across them.

Levelling

Where the subsoil levels are being adjusted or where you want a very level lawn the only way to be truly accurate is to use levelling tools. For large-scale work theodolites, quickset levels and similar tools are used, but for the average garden it should be perfectly satisfactory to use a spirit level, straight edge and wooden pegs. The alternative is to try to judge the levels by eye, a very risky business at the best of times unless you have had lots of practice.

Very few gardeners ever try to produce absolutely accurate levels but for the first-rate lawn it should be regarded as essential.

To set out pegs and level the lawn area you will need to obtain the following pieces of equipment:

a a straight-edged board at least 2 m long;
b a large spirit level;
c a tape measure;
d a number of wooden pegs which can be driven into the ground;
e a mallet.

You will need to select a datum point, i.e. a point from which all the levels will be taken. In most gardens you can do this by choosing a spot in the lawn area which is at the correct level already.

Step 1. Drive a peg into the ground at the datum point which you have just selected. The top of the peg should be about 10 cm above the soil level. A very accurate technique is to mark every peg as in Figure 4.3.

Step 2. Position the other pegs so that they form a square grid system with each peg 2 m from each of its neighbours (see Figure 4.4).

Step 3. Adjust the height of each peg by tapping it gently into the ground until the top of the peg is level with the top of the first peg that you installed. Use the straight edge and spirit level to check when the top of each peg is at the right height (see Figure 4.5).

With each peg marked as described here you can use the lines on the pegs to indicate whether or not soil has to be added to the spot. Tie a piece of string on to each peg at the marked line and run the string to the next peg. If the soil surface is exactly level the string will just touch the soil between the pegs; if the soil is too low or too high between pegs the string will indicate this. Adjust the levels by adding or removing soil as necessary (see Figure 4.6).

If you have used a spirit level and pegs to level the subsoil surface of your lawn site you can leave the pegs in position while you replace the topsoil. If your pegs are positioned so that the tops are 10 cm above the subsoil you should now add topsoil until it reaches the level of the peg tops. You will then have the minimum depth of soil and a level surface. You must alter the technique to suit your lawn of course. If you intend having 15 cm of topsoil then ensure that the tops of the pegs are that distance above the subsoil (see Figure 4.7).

At this point you should remove the pegs.

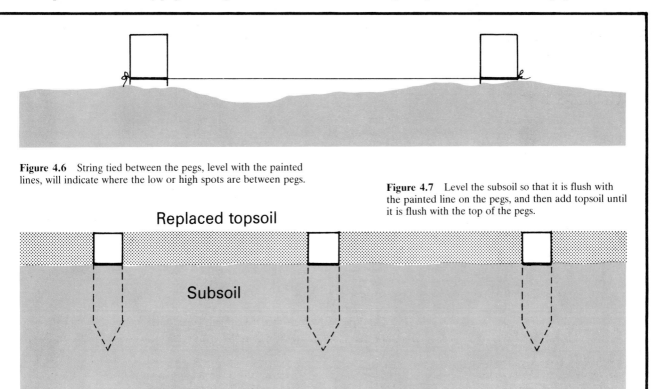

Figure 4.6 String tied between the pegs, level with the painted lines, will indicate where the low or high spots are between pegs.

Figure 4.7 Level the subsoil so that it is flush with the painted line on the pegs, and then add topsoil until it is flush with the top of the pegs.

Replaced topsoil

Subsoil

DRAINAGE

You have already been advised to check the quality of your soil and whether there are any potential drainage problems. The tell-tale signs of soils which have poor drainage are grey-blue colour in the soil, stagnant smells and the presense of standing water or plants which prefer wet soils.

Not everyone is prepared to go to the bother of installing drains and you should consider whether or not you will benefit from putting in such a system.

Small or large gardens can be drained in the same way. The methods and design of systems are identical whatever the scale of the work.

At what stage do I install drains?

During the construction of your lawn you have a choice of when to put the system in. The usual time is to put the drains into the ground after the topsoil has been returned to the site, but you might like to leave the system until after the grass has become established. Working on established grass is much cleaner.

To install drains in an established lawn area you should mark off the lines for each drain and carefully lift the turf. Remember that it has to be replaced. Dig out the soil and put it to one side of the drain trench. Some of this will be replaced in the trench but you should discard any subsoil which you dig out. Subsoil should not be put back into the trench because it will reduce the efficiency of the drainage system.

A drainage system is essential for first-rate lawns but for others you may find that it is easier just to add sand to the soil to improve its permeability. If there is no drainage system in your lawn you will have to accept that it will sometimes be too wet to walk on, but that shouldn't last too long. Regular spiking, as described in the aeration section of Chapter 7, will help to maintain good drainage.

If you have decided to put a drainage system into your lawn you again have a choice to make.

a Small sites may be drained perfectly well by building a soak-away, but

b larger gardens may need to have a full piped system.

Piped systems of drainage are expensive to put into lawns. So long as you are prepared to accept that your site will occasionally be wet, then you might never need to put in drains. If you want a lawn which drains well because you want to use it regularly, or because you want to practise your golf, tennis or bowls on it, then an efficient drainage system is necessary.

1 Soak-aways

A soak-away is simply a well in reverse. Water reaches the soak-away and then slowly seeps into the soil below. If the site slopes you should position the soak-away in the lowest part of the garden. Water will then move towards it. To build a soak-away:

a dig a pit about 1 m long, wide and deep;

b fill the pit with gravel or rubble to 30 cm below the top;

c add a 15 cm layer of fine gravel and then a 15 cm layer of topsoil to make the top of the pit level with the surrounding soil.

A soak-away is shown diagrammatically in Figure 4.8.

You could improve the efficiency of the soak-away by laying pipe drains in the garden and running them to the soak-away.

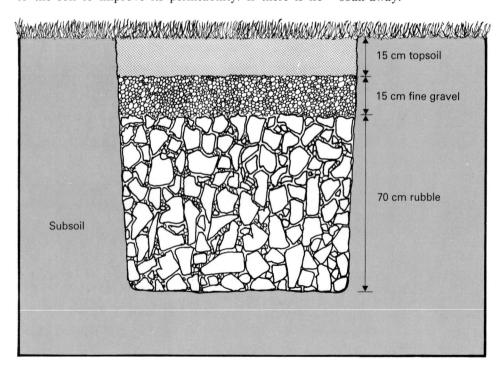

15 cm topsoil

15 cm fine gravel

70 cm rubble

Subsoil

Figure 4.8 Soak-away details.

2 Piped systems

Very few gardens warrant the expense and work involved in installing a system of pipe drains. Occasionally, though, you might come across a garden where the soil is so impervious that pipes do become necessary. They are, after all, the best method of removing water from very wet sites.

Are pipes really necessary in your garden?
Advising about drainage through the pages of a book is a little risky because no site can be known in detail. You must make the decision, after all you will be doing the work and paying the bills. If your soil is a heavy clay type and water stands on the surface of the soil for long periods you probably have a site which would benefit from having drainage pipes installed.

How do you know where and how to lay drainage pipes?
Drainage systems are normally laid in set patterns, the most popular being the herringbone and grid systems (see Figure 4.9).

Other layouts do exist and generally they are chosen to fit a particular shape of garden. The fan system is often used but many gardens can be drained perfectly well by running one pipe through the centre from the highest to the lowest point. A soak-away may also be installed at the lowest point to collect the water.

The pipes used may be made of clay or plastic. Both work perfectly well but plastics have the obvious advantage of being very light and easy to handle. Pipe size is determined by the size of the garden, although all but the largest of gardens should be adequately drained by installing 10 cm main drains and 7.5 cm laterals.

How far apart should the trenches be?
The distance between lateral drains should be determined by the soil type. Clay soils should have the drains fairly close together whilst in sandy soils they can be far apart if they are used at all.
– In clay soils the laterals should be 5 m apart.
– In medium loam soils the laterals should be 7–10 m apart.
– In sandy soils the pipes can be 15–20 m apart.
The pipes must be laid in trenches which are about 60 cm deep for laterals and 120 cm deep for main drains. Keep the trenches as narrow as possible to reduce the amount of back-fill used. They should have a consistent slope or fall of between 1 in 80 and 1 in 200 and there should be no sudden changes in level or direction of pipes.

The main drain should run down the slope of the garden. If the site is level you will have to create a fall in the trench itself. Line the base of the trench with gravel before laying the pipes.

If you use clay pipes be sure that you butt the individual pipes as close together as possible.

Cover the pipes with gravel of size 6–9 mm to a depth

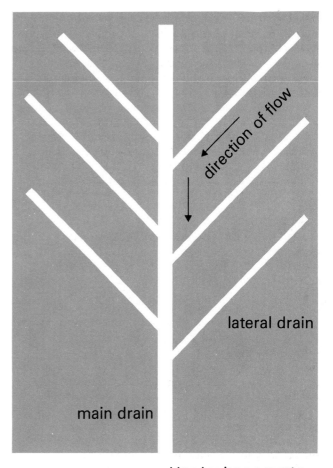

a Herringbone system

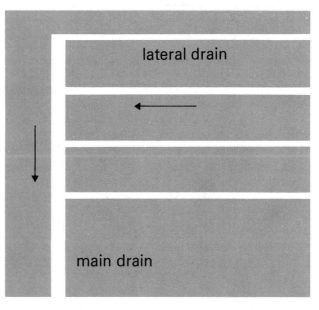

b Grid system

Figure 4.9 Drainage system patterns.

of approximately 30 cm and then cover this with a 5 cm layer of coarse sand. Finally fill up the trench with topsoil to which sand has been added. Discard the subsoil; if you put it back in the drain trench it will only reduce the efficiency of your new system. Sand is added to the topsoil so that water can move through it more easily. The details are summarised in Figure 4.10.

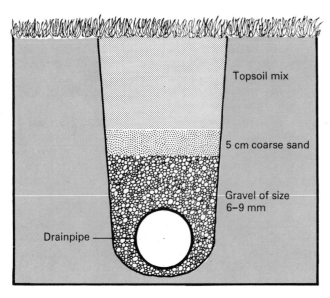

Figure 4.10 Drain backfill details.

When can I install a piped system?
Drainage pipes should be installed only after the site has been levelled. It may be a lot easier to wait until the grass is fully established and then you can work in fairly clean conditions. If your garden is so large that it is not possible to dig all the trenches by hand you might like to consider hiring a trencher, a machine which will dig very clean trenches for you. Trenching machines certainly save a lot of hard work and they can be used safely after only a few minutes' training.

CULTIVATING AND IMPROVING THE SOIL
The fourth stage of the preparation work provides you with the opportunity to break up the soil and also to improve it in whatever ways are necessary. We have already mentioned drainage systems but some soils may be drained by adding sand. Other soils may be so droughty that they have to be improved to make them able to hold water. The final type of improvement which you might need to carry out includes the addition of fertilisers.

The soil must be thoroughly cultivated to a depth of 20 cm unless the topsoil layer is shallower than this (see Figure 4.11). Avoid mixing top- and sub-soil layers.

Digging is the most common method of cultivation but you may wish to use a mechanical cultivator to save time and effort. If you do wish to use a machine to do the cultivation work avoid creating too fine a tilth which will cap or compact easily. You must also remember that when a rotary cultivator has been used it tends to leave a very loose and fluffy seed-bed which must be firmed evenly before seed-sowing or turf-laying can begin.

If you cultivate the site during the autumn and the soil is very heavy aim to leave the ground in a very rough condition during the winter to let the frost break down the large clods for you.

While you are cultivating the soil take the opportunity to improve it. pH is a measure of the acidity of the soil. A pH too low or too high may encourage weeds and diseases to appear in the lawn; it may also weaken the grass plants.

What is the pH of your soil?
A simple soil-testing kit can provide the answer. If the pH is below 6.0 you should consider adding lime to raise it. The range that should be aimed at is pH 6.0–7.0 but if your soil is close to this range do not bother trying to change it. Grasses will tolerate a fairly wide range of pH so don't be too worried unless your soil is below, say, pH 5.0.

It may not be possible to do anything if the pH is above 8.0 because of the type of soil in the area and even the quality of the water. Peat or garden compost will help to lower the pH, but so too will some of the types of fertiliser which you will be using on the established lawn.

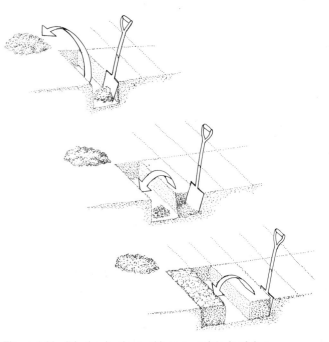

Figure 4.11 Dig the site thoroughly to a spade's depth but avoid mixing topsoil and subsoil layers.

Does your soil dry out too quickly?
Some soils, particularly the sandy types, dry out very fast indeed. The ability of the soil to hold water can be improved by adding organic matter to the soil in the form of peat or garden compost. Add about 3 kg compost to each square metre of soil surface.

Does your soil stay wet for too long after rainfall?
We have already mentioned installing drainage pipes. Many gardens do not need such elaborate treatment and their problems can be solved by improving the permeability of the surface soil by adding sand. About 13 kg sand per square metre will be needed but use a medium grade. Coarse sands do not improve the soil as much as medium sand because the spaces between the coarse sand particles are easily filled by the smaller soil particles. Medium sand has particles in the range of 0.1–0.6 mm and will provide a long lasting improvement to drainage. Use it at the rate just mentioned and thoroughly cultivate it into the top 10–15 cm of soil.

The addition of sand will undoubtedly help drainage but it will also help to maintain good aeration. This in turn will encourage the development of deep, healthy root systems and consequently a healthy lawn.

Are perennial weeds a problem?
Weeds such as couch grass and ground elder are a problem in many gardens. While you are digging the site pull out any perennial weeds. Weeds which are growing can be killed off by the use of chemicals.

FINAL PREPARATIONS
As you get nearer to the time for sowing seed or laying turf you should be trying to produce the perfect seed-bed, i.e. one which is firm and fine.

Rough cultivation work has left large clods of soil which must now be broken down to produce a fine tilth. On small sites use the back of a garden fork to smash the clods into smaller pieces. On larger sites use a rotary cultivator with the blades set to work very shallowly. The cultivator will create a shallow tilth but you must be careful not to reduce the surface to dust.

Now that the surface is fairly fine you must remove any large stones and other unwanted debris such as roots or twigs which might have been brought to the surface by digging. The easiest way to remove such debris is to pull it off using a rake: this cuts out the bending and is good practice for the following stages of the work.

What has been achieved so far?
The site for your lawn should now be level and smooth. You still have some work to do and the aim of the following operations is to produce the final firm tilth which will encourage quick establishment of the grass.

Firming the site
If you simply laid turf or sowed seed on to the site in its present state you would eventually have a lawn which settled very unevenly and which would be filled with hollows and hills. You must firm the soil evenly so that there is no risk of uneven settlement in future years.

Should I use a roller?
Briefly, the answer is 'no'. Treading the site is by far the better method because it enables you to pick out the soft areas and firm them carefully. A roller can miss soft spots if they happen to coincide with the centre of the roller. For maximum effect leave the site as long as possible between cultivating it and treading it. Nature will then do much of the work for you.

When you tread the site make sure that the soil is fairly dry or you will cause excessive compaction of the surface.

To tread a site you should walk over the soil with short, overlapping steps putting all your weight on your heels. Work systematically across the site until it has all been firmed.

Follow the treading by careful raking to produce a smooth surface, free of debris and large stones. If the site is still too rough repeat the operations of treading and raking until the correct tilth has been produced. The correct tilth is when the soil crumbs are about the size of grains of wheat and not much smaller.

Fertiliser treatment
Shortly before the great sowing or turfing day you should make the final improvements to the soil – adding fertiliser.

Why should seeds require fertiliser?
If your lawn is to establish quickly the grass plants must have ready access to a supply of food. If the grass establishes slowly you will find that weeds encroach and virtually take over before your seed has the chance to germinate. Produce a dense lawn quickly and you minimise the problems with weeds for the future.

Pre-seed fertiliser is that which is applied to the ground shortly before laying turf or sowing seed. This type of fertiliser should supply the three main nutrients, nitrogen, phosphates and potassium. Despite popular belief, phosphates on their own do not stimulate rooting; they must be in balance with the other two nutrients.

A useful pre-seed fertiliser can be made by applying the following chemicals about 7–10 days before sowing the seed:

20–25 g/m² superphosphate (18% phosphorus)
6–10 g/m² sulphate of potash (48% potassium)
6–10 g/m² sulphate of ammonia (21% nitrogen)
or you can use any proprietary pre-seed fertiliser for lawns.

Spread the fertiliser evenly by hand or mechanical spreader. For accuracy you can divide the site into strips using string or garden lines. Each strip should be about

1 m wide. Rake the fertiliser into the ground and leave it for a few days before you sow the seed or lay the turf. The time interval between fertilising and seeding the lawn should allow the fertiliser to dissolve and become available for plant use.

After these preparations the site will be ready to accept the grass: it should be weed-free, drained, firm and level. If the soil is moist grass establishment should be fast.

Prepare a time chart, like the one given in Table 4.2, for the work on your lawn.

CHECKPOINT
Try to answer the following questions without referring to the chapter.
1 What is meant by major grading?
2 Is major grading necessary in your garden?
3 What fertilisers can be used to prepare a pre-seed fertiliser?
4 What is the sequence of operations required to prepare a seed-bed?

Now look back to check your answers.

Table 4.2 **Work time chart**

Operation	Suggested time	Your timing
Site clearance	Spring	
Grading and levelling	Early summer	
Drainage	Summer	
Cultivating the site	Summer	
Fallowing the site	Summer	
Treading and raking the site	Summer	
Sowing or turfing the lawn	Autumn	

The lawn is an essential feature in most gardens (above), providing a play area and a backcloth for other plants.

Avoid creating narrow strips of grass which are fussy to the eye and difficult to mow.

Trees with dense canopies will cast heavy shade on lawns, killing out the grass and making access with the mower difficult.

The ideal lawn; neat, level, dense and weed free.

Modern turf production may involve growing the specially sown turf on beds of concrete or tarmacadam. This system produces uniform turves of equal thickness and composition.

Specialist machinery is employed to cut and lift quality turf in the turf nursery (right).

After lifting the turf is stacked ready for immediate transportation to the site where it can be laid without delay (above).

Provided the ground is well prepared and levelled, the laying of evenly cut turves is child's play.

The great advantage of turf is its instant effect. What was a bank of soil in the morning can be a lawn in the afternoon.

Irrigation (above) is essential in dry weather to prevent the grass from browning, but it is vital that the earth is thoroughly soaked. Garden sprinklers (above right) are far better than cans and hoses for watering lawns. Left running on one patch of grass for several hours they will apply water so that it is evenly absorbed by the lawn, leaving you to get on with other jobs.

Fertiliser must be evenly spread to avoid scorching the grass. A wheeled spreader (below left) will accurately deliver the correct amount of food to the lawn.
Hollow tine aeration can stimulate deeper rooting as well as improving surface drainage. These healthy new roots (below) are growing in the hole left by an aerating fork.

Clippings left on the surface of the lawn (above) can look unsightly and, if present in large amounts, they will kill out the grass and encourage disease. Always be on the lookout for tell-tale signs of surface compaction leading to poor drainage (below). Take remedial action quickly to safeguard the grass.

Fusarium patch – the most common disease of fine lawns.

Fairy rings (left) – showing the three grades or types of fairy ring.

Some diseases such as this one (below) – ophiobolus patch – are encouraged by excessive use of lime.

Red thread disease (*Corticium fuciforme*) is easy to diagnose once the pink needles are visible (below left).

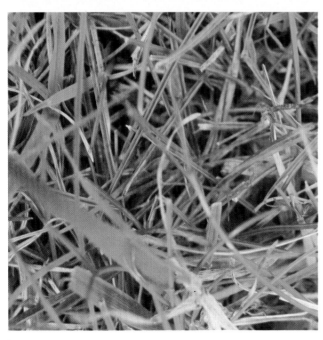

Not all brown patches are caused by diseases: bitch urine may be to blame or, in this case, the careless use of herbicide.

Drought damage is usually most severe on steep banks where the drainage of water is rapid.

Healthy lawns, like this one, may look pale during severe drought but soon recover when rain falls again.

Scarify your lawn at least once a year with a spring tine rake to remove dead grass or 'thatch'.
Aerator rakes (below) remove weeds and dead, matted grasses.

Leaf sweepers (left) make light work of clearing both clippings and leaves from lawns. A rotating brush flicks the debris into a collection bag.
Powered lawn rakers (below) make the job of scarifying a large lawn much easier.

39

Edging shears are the most useful tool for lawn edging, but wheeled cutters are speedier.

Cylinder mowers produce the finest cut and the finest lawns. No other mower will create that velvety finish with neat stripes.

Rotary mowers may not be suitable for fine lawns but they are by far the best choice for an average hard-wearing lawn. They can be bought fitted with grass boxes to collect the clippings.

CHAPTER FIVE

Which grass?

Glossary of terms used in grass identification
Choosing grasses/Seed identification
Turf for lawns/Checkpoint

There are more than 10,000 different species of grass in the world. In Britain there are about 150 species, yet for lawns and grass areas which are used for sport we use only about 15 of these species. What is it that makes some grasses suitable for use in lawns yet makes others unsuitable?

Try to make up a list of the qualities which you would like your grass plants to have. These qualities should make the grass perfect for the use you have in mind.

The ideal grass for lawns should be:

a quick to establish from seed;
b able to spread vegetatively by stolons or rhizomes (see glossary of terms below);
c able to produce a lawn which looks good throughout the year;
d disease- and pest-resistant;
e tolerant of wear;
f quick to recover after heavy wear or damage;
g able to produce a dense turf;
h slow-growing to reduce the need for mowing;
i easy to cut, leaving no flower spikes uncut;
j relatively free from fibre.

Before we go on to look at individual grasses remind yourself about any potential problem areas which exist on your lawn site. Are there any shady spots or very wet spots? Problem areas must be identified if you are to be able to select the right grasses for your lawn.

Remind yourself of the answers you gave to these questions from Chapters 1 and 2: Do you want a fine lawn, or a hard-wearing lawn for the children to play on? Do you have the time to maintain a very fine lawn? Do you want to make your lawn from seed or turf?

The glossary below explains the terms used in the rest of this chapter. Some of these will be new to you and it is important that you understand each one.

Glossary of terms used in grass identification

Leaf blade
The free part of the grass leaf.

Leaf sheath
The part of the leaf which clasps the stem.

Leaf blade
Leaf sheath

Rhizome
A creeping stem found growing just below the soil surface.

Stolon
A creeping stem found growing on the soil surface.

Auricles
Small claw-like structures found at the junction of the leaf blade and sheath.

Ligule
Auricle

Ribbed leaf

Rib

Non-ribbed leaf

Tramlines
Depressions found running the entire length of meadow grass (*Poa*) leaves.

Tramlines

Boat-shaped leaf tip

Rolled leaf

Folded leaf

CHOOSING GRASSES

To choose the right grass you need to be fully aware of the difficulties which your lawn might have to face. You should also know whether you want fine grasses, hard-wearing grasses or grasses that will tolerate neglect.

What kinds of grass can I choose from?

The descriptions of grasses which follow should answer your questions and enable you to choose those grasses which are most suitable for the type of lawn you want and any special conditions that prevail.

For fine lawns

The grasses which are used for fine lawns must be able to tolerate very close mowing and they must have fine leaves with good colour which are densely packed to produce a high-quality lawn.

For fine lawns you are restricted to two distinct genera or groups of grasses, the bent grasses (*Agrostis*) and the fescues (*Festuca*). These two groups of grasses form the basis for most fine lawns and are also used to thicken up the sward in more hard-wearing areas.

Browntop bent (*Agrostis tenuis*) is probably the most common bent grass in lawns. It is a tufted perennial which is found naturally on heaths, moorland and waste ground, especially where the soils are rather dry and acid.

Browntop bent may have short stolons or rhizomes, it is often slow to establish but if regularly mown it can produce a very dense turf of very high quality.

The leaf of browntop is fine and tapers to a fine point along the entire length of the leaf blade. The surface is lightly ribbed and the leaf will become finer as the frequency of mowing is increased.

Creeping bent (*Agrostis stolonifera*) is less common than browntop but can also produce very fine lawns. This grass produces vigorous stolons and if allowed to grow unhindered it will produce a very spongy lawn. It must also be raked regularly if it is to be at its best. Creeping bent does not enjoy badly drained soils, especially in the winter, nor will it do well if the soil is ever subjected to summer drought. This is a hard-wearing grass but it needs a very high standard of maintenance to keep it looking good.

Velvet bent (*Agrostis canina*) makes a very attractive sward but is no use at all for areas which must take a lot of wear and tear and is not a common grass in lawns. This grass would be excellent for damp, shady spots such as pond margins but would quickly die if grown and subjected to drought. Velvet bent spreads by stolons but is easily torn out by a mower which is set to cut too low. The leaves are soft and velvety.

Figure 5.1 Grasses may be identified by their ligules.

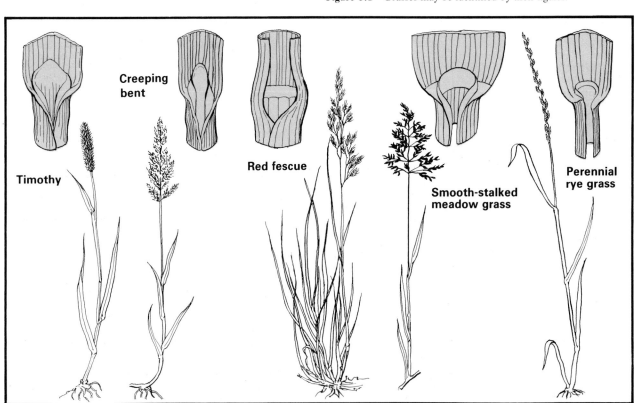

Red fescues are a group of grasses that have been sub-divided to form smaller groups of very useful fine-leaved grasses.

Chewing's fescue (*Festuca rubr* ssp. *commutata*) is perhaps the most well known of these sub-groups. This grass is often used along with browntop bent to form the basis of fine turf areas. Chewing's is a very fine grass with needle-like leaves. It is ideally suited to light soils and is very tolerant of dry conditions. This grass does not have rhizomes but it tillers, i.e. produces side shoots, very well and so produces a dense lawn.

Creeping red fescue (*Festuca rubra* ssp. *litoralis*) has short slender rhizomes and again produces very fine, needle-like leaves. It is very similar to Chewing's fescue but often survives drought much better. Creeping red fescue is much more wear-tolerant than Chewing's but does not tolerate close mowing quite so well.

There is another form of creeping red fescue, *Festuca rubra* ssp. *rubra*, but this is not used quite so much as it forms a coarser and more open lawn than the slender type of creeping red described above.

Sheep's fescue (*Festuca ovina*) is not often used for amenity lawns but may be of value if you have an area where you would like to create a natural appearance without mowing. Sheep's fescue produces very fine leaves, again like needles, but grows in distinct hummocks.

There are other forms of fescue but they are of little value and so I will not describe them here.

For coarse or hard-wearing lawns

In the main this type of lawn will still contain the fine-leaved bents and fescues which make up the finest of lawns but they will also contain coarser and stronger species of grass which tolerate more wear.

The main coarse grasses are to be found in three plant genera, perennial rye grass, Timothy and the meadow grasses.

Perennial rye grass (*Lolium perenne*) is, as its name suggests, a perennial. It is the most common grass in hard-wearing areas and is characterised by very fast establishment and growth. Like the other coarse grasses, *Lolium perenne* requires rather more feeding and watering than the very fine grasses. It also prefers a higher pH. *Lolium* is a tufted grass which does not spread vegetatively; it is normally mixed with fine grasses to thicken up the sward. Perennial rye grass can be easily recognised in lawns because of the very shiny underside of the leaf, the dull ribbed upper surface and the red base of the shoot. It is also one of the few grasses to have ligules at each leaf junction. There are several new cultivars or varieties of this grass produced each year, each one having finer leaves and requiring less cutting than the older cultivars. Perennial rye forms the basis of hard-wearing lawns but it requires a lot of mowing because it grows so quickly.

The leaves of this grass may turn brown at the tips in autumn. Drought may produce similar effects.

Timothy (*Phleum bertolonii*). This is an improved form of the old agricultural Timothy but it has similar characteristics. It does well on heavy, wet soils and has very attractive greyish foliage. At the base of each shoot you should be able to find a very distinct swelling, rather like a spring onion. Do not use this grass on dry, shallow soils. It provides a nice colour contrast with other grasses in the right soils.

Smooth-stalked meadow grass (*Poa pratensis*). The Americans call this Kentucky blue because of its leaf colour. It is an attractive grass, second only to perennial rye in wear-tolerance. Like all meadow grasses this one has two very distinct lines, referred to as tramlines, running down the centre of each leaf. Each leaf tip is boat-shaped, and most of the young leaves have crinkles in them. The shoot of meadow grass is flat because in the young stage each leaf is folded whereas grasses such as bents have rolled leaves.

Recommended grass varieties

Grasses, like vegetables and flowers, have cultivars or varieties. There are many from which you might choose. A few of the better cultivars are listed in Table 5.1.

Table 5.1 **Recommended cultivars**

Grass	Recommended cultivars
Perennial rye grass	Manhattan, Sprinter, Majestic, Derby
Timothy	Nobis, S50, Ramona
Smooth-stalked meadow grass	Parade, Baron, Bensun
Chewing's fescue	Frida, Highlight
Creeping red fescue	Dawson, Merlin
Browntop bent	Highland, Bardot, Tracenta
Creeping bent	Penncross, Penneagle
Velvet bent	Kingstown

Special considerations

What else should I consider when I choose my grass?
You must take into account the pH of your soil as grasses have preferred soil pH levels (see Table 5.2).

If your lawn site has special needs Tables 5.3 and 5.4 will help you to choose the right grass and seed mixture for it.

If your garden presents unusual problems you may need to modify the seed mixture slightly. Your seed supplier should be able to advise you on this.

Grass	Preferred pH
Perennial rye grass	6.0 — 7.0
Smooth-stalked meadow grass	6.0 — 7.0
Timothy	6.0 — 7.0
Browntop bent	5.5 — 6.5
Creeping bent	5.4 — 6.5
Velvet bent	5.0 — 6.0
Sheep's fescue	4.5 — 5.5
Red fescues	4.3 — 5.5

Table 5.2 **Preferred pH levels for grasses**

Grass	Tolerant of		
	salt	shade	heavy wear
Creeping bent	Yes	No	No
Creeping red fescue	Yes	No	No
Velvet bent	No	Yes	No
Perennial rye grass	No	No	Yes
Smooth-stalked meadow grass	No	No	Yes
Timothy	No	No	Yes
Browntop bent	No	No	No
Chewing's fescue	No	No	No

Table 5.3 **Grass tolerance**

Table 5.4 **Seed mixtures**

Seed mixture (parts by weight)	SITE
80% Chewing's fescue 20% Browntop bent	Fine ornamental lawns cut at 5—6 mm.
40% Chewing's fescue 40% Creeping red fescue 20% Browntop bent	Fine lawns cut at 10 mm.
35% Creeping red fescue 30% Chewing's fescue 25% Smooth-stalked meadow grass 10% Browntop bent	Lightly shaded areas where the grass can be allowed to grow long.
40% Chewing's fescue 30% Perennial rye grass 20% Creeping red fescue 10% Browntop bent	Hard-wearing lawns.
45% Creeping red fescue 30% Chewing's fescue 20% Browntop bent 5% Timothy	Banks and steep slopes.

SEED IDENTIFICATION

Grass seeds are very distinctive and fairly easy to distinguish. Table 5.5 giving seed size can be related to the diagrams in Figure 5.2. In general the larger the seed, the more vigorous will be the plants; thus if you come across very small seed you will find that it belongs to one of the finer lawn grasses.

Grass	Approximate number of seeds per gramme
Browntop bent	15,000
Timothy	4,000
Smooth-stalked meadow grass	3,000
Red fescues	1,000
Perennial rye grass	600

Table 5.5 **Seed size**

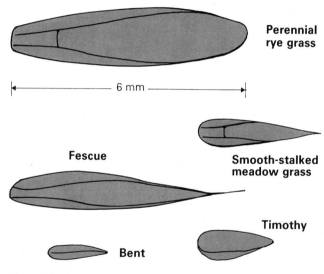

Figure 5.2 Seed identification by shape.

Perennial rye grass is obviously the largest of the seeds we deal with. It is boat-shaped and stands out in most mixtures.

Fescue seed should always reach you with a spike still intact on each seed. This spike is called an awn and should be easy to find.

Smooth-stalked meadow grass has quite small seeds, which, if rolled between the finger and thumb, feel angular. They are actually triangular.

Timothy has small, very distinctive, white seed.

Bent grasses have very small seed, by far the smallest of all. In a mixture this seed tends to fall to the bottom of the bag and often sticks to your palm at the time of sowing.

TURF FOR LAWNS

We have already considered the advantages and disadvantages of using turf. One of the main problems with buying turf is the difficulty in obtaining good quality supplies.

If you buy turf from a supplier, whatever type of turf it might be, you should always check it for weeds and signs of disease. There should be no bare patches and the soil on which the turf is grown should be stone-free.

Traditional types of turf

Perhaps the most common, and certainly the cheapest, form of turf is meadow turf. This type of turf may be green and cheap but it is far from ideal for lawns because it contains vigorous and coarse agricultural grasses rather than the finer lawn types.

Where would you expect to get the finest type of turf?

I'm sure that you have been to seaside areas or moorland areas where the grass seems to be very fine indeed. Both areas can produce good quality turf but both may also introduce their own problems.

Moorland turf is often very fine but the grasses tend to be matted and there is often a fairly thick thatch layer already present.

Turf grown near the seashore has been famous for many years as the finest available. Sea-washed, or Cumberland turf as it is often known, is composed of fine creeping red fescue and creeping bent but the soil on which it is grown is very silty. You will find the silty soil very difficult to manage and the turf will change in character within a few years.

The cheapest turf available, as described above, comes from farm fields which are stripped of their grass cover. Occasionally the turf will have been cut regularly to give it a neater appearance and many producers will also weed-treat the turf before it is sold. No matter how much care has been taken, the turf will still be pasture turf and it will still contain the wrong grasses for good quality lawns. If your aim is just to create something green then this type of turf will certainly do that.

Downland turf is also fine in its texture but, unlike the moorland turf, is not usually matted or full of thatch. Quality is generally good and this turf constitutes a good buy.

Seedling turf products

The types of turf described so far have all been those which are grown for several years before being sold. Within recent years a new approach to turf production has been introduced – seedling turf products. There are three such products: SAI Turf Mat, Bravura Turf and Tana Turf.

SAI Turf Mat is now sold by Rolawn Turf Producers in York. The turf is roughly seven weeks old when sold and will have been grown on a peat bed so that the turf comes to you weed-free and vigorous. Many garden centres now stock Rolawn turf.

Bravura turf is usually about six to seven weeks old when sold and will be growing on a weed-free base which is reinforced by a type of plastic mesh. Details may be obtained from Bravura Ltd of Weston-super-mare, Avon BS24 9UB.

Tana turf is about the same age as the others but is grown in a very different way. It is produced on a bed of water, a type of hydroponics. Details may be obtained from Plantagenet Seeds of Pickering in North Yorkshire.

Each of the seedling turf types requires careful management, especially during the early days when it is just becoming established. There is the great advantage with each of the three methods that the seed mixture can be specified by the buyer. The turf comes neatly rolled and weed-free and is a far cry from many of the supplies sold in the past.

Semi-mature turf

Perhaps the best turf on the market, semi-mature turf, comes from Rolawn. The turf is specially grown for landscaping use and is made up entirely of horticultural grasses rather than agricultural types. The turf is grown for about 12 months before it is sold. The 12-month period is important because that is the minimum time for turf roots to establish fully and bind the soil together. The turf is sold in rolls, each roll covering 1 m², so that laying is easy and fairly quick.

Modern techniques of turf production have therefore produced a much better product than was ever available before. It is, however, rather more expensive than the meadow or pasture turf which is so readily available.

CHECKPOINT

Try to answer the following questions without referring to the chapter.

1 What should you look for when selecting turf for a good quality lawn?
2 Which grasses should form the main part of a seed mixture for a very hard-wearing lawn?
3 What is the difference between a stolon and a rhizome?
4 What types of turf are currently available and which type would you select for a very fine lawn, and for a low-grade lawn?

Now look back to check your answers.

CHAPTER SIX

Establishing the lawn

In Chapter 2 you considered the reasons for choosing either seed or turf to create your lawn. If you have not yet decided which method to use Table 6.1 provides a summary of the advantages and disadvantages of each. You can always refer back to Chapter 2 but you should be able to choose to use seed or turf on the basis of this chart.

I will cover the two methods of establishing lawns in turn.

LAWNS FROM SEED

Here are some questions you might ask about creating a lawn from seed:

What is the best time to sow grass seed and why?
How much seed should be sown to establish a lawn?
Should the seed be covered after sowing?
What are the seeds' worst enemies?

When to sow

I will begin by looking at the best time to sow seeds. The ideal soil conditions for sowing are those which supply the germinating seed with air, moisture and warmth.

If the seed-bed has been well prepared it should be well drained and will therefore have a good supply of air, and oxygen in particular. Of course you can always water the site to supply water but nature usually does this quite well and at the same time ensures that there is enough warmth for plant growth. The best times during the year for seed-sowing would seem to be late summer – during late August and early September – and spring – April and May. At both these times the soil should still be warm and moist.

Given a choice I would always aim to sow grass seed during late August because a spring sowing is always at risk from spring or summer droughts. In a good seed-bed an August/September sowing should be well established before the winter sets in and slows down plant growth.

Having mentioned the best times to sow grass seed it must be said that summer sowing can be very successful if the weather is kind and irrigation is always available. I have even been successful with sowings made as late as the middle of November, but again the weather can so easily determine success or failure.

The speed of growth of grasses will determine how quickly they cover the soil and also whether they allow weeds to come into the seed-bed or not. Large seed tends to grow more quickly than small seed, which may take as much as two or three times as long to germinate and emerge. Emergence is the term given to the appearance of the grass on the soil surface: germination has obviously already occurred by the time of emergence.

Table 6.1 **Advantages and disadvantages of turfed and seeded lawns**

Turfed lawns	Seeded lawns
Quick effect; instant lawn.	Slow to establish.
Can be used almost immediately.	Needs to be allowed to settle in.
Quite expensive.	Cheapest.
May bring weeds in the soil.	No weed problem.
Diseases and pests may be in the soil.	Diseases may attack the young grass.
Turfing can be heavy, skilled work.	Less skilled work is required.
Often difficult to get a good supply of turf.	Good supplies are easy to get.
Turf cannot be stored for long periods.	Seed can easily be stored for long periods.

Table 6.2 provides details of grass emergence and is a useful reference guide. It also gives details of the time taken for the grass to grow to a height of 5 cm, the stage at which the first cut should be made. If you are sowing perennial rye grass, for example, don't plan to go away on holiday three weeks later because when you return your grass will be too long for easy cutting.

Table 6.2 **Rate of seed emergence of lawn grasses**

Grass	Days from sowing to emergence	Days to reach 5 cm in height
Perennial rye grass	7	18
Timothy	8	26
Bent grass	12	32
Smooth-stalked meadow grass	14	34
Red fescues	10	30

＊ *The figures were obtained in good growing conditions. If the weather is poor or the weather conditions adverse the times for each stage will be longer.*

How much seed do you need?

Before you can achieve anything you must know how much seed you need. To find this out you need to
a work out the area of your lawn;
b know what kind of lawn you want.

Check the measurements of your lawn and from these work out the total area of ground which is to be sown with grass seed. For example, if your lawn measures 6 m by 10 m, the area will be 6×10 m^2 = 60 m^2.

If your lawn is an odd shape you may need to divide it into several smaller sections. The area of each section can then be calculated. An example of this is given in Figure 6.1.

If you drew your first plan on graph paper it should be fairly easy to work out the area of the lawn by simply counting the squares.

Your answer for the area of the lawn may come to an odd figure. In this case you should always round it up a little to ensure that you have enough seed for the work.

Which type of lawn did you decide to create?

This question is important because it determines how much seed you need to buy of a particular type of grass.

If you chose to produce the finest lawn possible – say an 80:20 Chewing's fescue and browntop bent mixture – you should be sowing it at a rate of 35 g/m^2. If you chose a mixture with a lot of larger seed in it – say extra red fescue or perennial rye grass – then you should be sowing this mixture at a rate of 35–50 g/m^2.

Don't assume that if you double the seed rate you will improve the lawn; in fact the opposite is true because you will cause overcrowding and that is enough to encourage diseases such as damping off and fusarium. These and other diseases are described in Chapter 8, 'Lawn problems'.

The next stage in your work is to calculate how much seed to buy. Here is an example:
– Lawn to be 6 m by 10 m. Quality: very fine. Rate of sowing 35 g/m^2.
– Seed required: $6 \times 10 \times 35 = 2,100$ g or 2.1 kg. You should buy 2.5 kg seed.

When it comes to working out the final quantities of seed for the lawn you should always be a little on the generous side. Allow some for the lawn, some for the birds and some to fill up the bare patches which you forget to sow. As a rough guide you should allow for an extra 10 per cent seed. Seriously, though, seed is fairly cheap and you should always buy a little extra just in case patches do not germinate as you would wish.

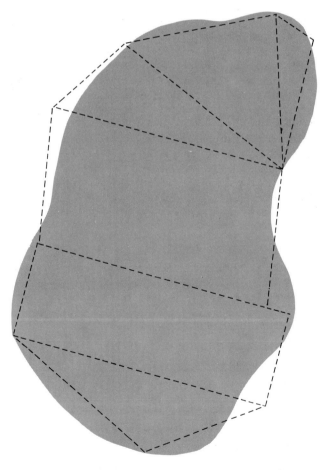

Figure 6.1 Divide irregular lawns into regular shapes to make your area calculations easier.

Sowing the seed

Having bought the seed, the next step is of course to use it. If you have a good seed or fertiliser spreader then this will make light work of seed-sowing, but it must be calibrated or set up to sow the right amount of seed. Few people will have such a piece of equipment and I'm sure that nearly every lawn has been sown by hand.

You need to spread the seed evenly and thinly for maximum effect. To do this should follow the instructions below.

a Measure out enough seed to sow 1 m².

b Put the seed into a paper cup or plastic margarine container if it's a small one.

c Cut the cup down in size so that it is filled completely by the seed. You now have a useful measure which will measure out just enough seed for each square metre of lawn area.

d Divide the lawn area into squares using string or garden lines as shown in Figure 6.2. Each square should be 1 m × 1 m.

e Using your measure, scoop up enough seed to cover one of the squares.

f Divide the seed into two lots, roughly equal in size.

g Sow one half of the seed into the first square by scattering the seed gently as your hand moves left to right. Scatter the other half of the seed by moving your hand up and down at right angles to the first direction of sowing, as shown in Figure 6.3.

h Repeat this operation on every square until the lawn is completely sown.

i Remove the string or garden lines which marked the lawn into squares, and then lightly rake the surface so that most of the seed is covered. Grass seed does not have to be covered in order to grow; some of the finer seeds do better if left on the surface rather than covered by soil. Rake to leave very shallow furrows in the soil and do not cover even the larger seed by more than 6 mm.

Protection from birds

Your main enemy will no doubt be birds – those that eat the seed and those that come to dust-bathe in your nicely raked seed-bed.

Some seed is already treated against birds and will discourage their depredations, although I should stress that I have sown many acres of grass without ever using treated seed. If your seed-bed is well made and you sow the correct amount of seed there should be enough for the lawn while allowing for some robbery by the birds.

Birds which dust-bathe in your seed-bed can be an absolute menace. The way to stop this is to keep the seed-bed evenly moist in dry weather; turn on a sprinkler for a few hours. Alternatively you can spread black cotton across the lawn so that it criss-crosses the site about 10 cm above the soil, but birds can be crippled by black cotton which winds around their legs. It's safer to cover small sites with plastic netting.

Germination

After sowing, the seed must have water, air and warmth in order to germinate. Your main concern will no doubt be the water supply. If dry weather occurs shortly after sowing, especially with light soils, you should be ready to irrigate to keep the seed and soil moist, not wet. Once the seedlings appear above the soil surface you should reduce the amount of water given to the grass. Less water at the surface means that the plants must send their roots down in search of water. You must encourage deep roots; regular irrigation would ensure shallow roots and perhaps problems in the future.

Some gardeners like to roll their seed-beds. If you trampled the site well and if the seed was carefully raked into the ground you will rarely, if ever, need to roll the soil. More often than not rolling causes future problems.

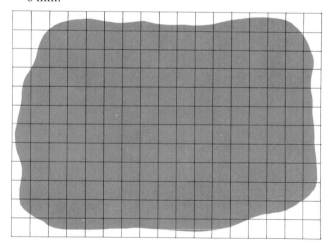

Figure 6.2 Divide the lawn into 1 m wide strips and then 1 m squares using string lines.

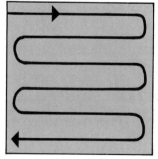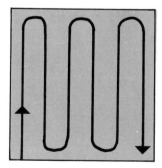

Figure 6.3 Sowing the seed. Half the seed is scattered as your hand moves left to right in a zig-zag pattern. The second half of the seed is then sown at right angles, again in a zig-zag pattern.

After germination occurs, what then?

I mentioned that rolling was often harmful on seed-beds. On some soils (the light ones) rolling can be useful after emergence of the seedlings. A light rolling will settle the soil around the roots of the new seedlings and will help them to take up water from the soil. Rolling will also encourage the young plants to tiller, that is to produce side shoots. The greater the number of side shoots, the denser will be the lawn. A light rolling will also press stones into the surface so that they do not interfere with the blades of the mower.

First mowing

At the beginning of this chapter Table 6.2 gives details of the rates of growth of some of the more common grasses. Refer to this chart now for guidance about when to start mowing.

The young grass should be given its first cut when it is between 35 and 50 mm high. The blades of the mower should be sharp and set quite high so that they remove no more than one-third of the grass foliage. If you let the grass reach, say, 45 mm in height it can be cut to leave 30 mm of foliage. Removing more than this can weaken the plants too much.

What should I use to cut the grass?

The ideal mower would be the hover type for the first cut. If you have a cylinder mower make sure that the blades are well set and very sharp before cutting the grass. If the mower blades are not sharp there is a risk that they will pull the young grass plants out of the soil.

Light mowing is essential in the first few weeks of the new lawn because it will encourage deep roots and strong tillers.

How often should I cut the new lawn?

Use the one-third rule and never remove more than this amount of the foliage. As the lawn cover becomes more dense the grass should become a little more vigorous and if the seed was sown in spring the height of cut can slowly be reduced at each mowing. With an autumn sowing leave the grass at a height of 35 mm for the winter and reduce the height of cut next spring.

What else should I do to the new lawn?

Undoubtedly there will be weeds appearing with the new grass. These must be dealt with now if they are not to become greater problems in the future.

Weeds may be grasses or broad-leaved types. Weed grasses such as the grey-leaved Yorkshire fog (*Holcus lanatus*) can be easily spotted and must be pulled out by hand because they cannot be killed off to leave the desirable grasses alive. Broad-leaved weeds can be treated with chemicals but not until the grass has matured or has two leaves. At this two-leaf stage only special chemicals should be used. Most selective weedkillers for lawns can-

not be used until the grass is at least six months old. Suitable chemicals for seedling lawns include ioxynil. Annual weeds such as chickweed and groundsel will be killed off by regular mowing.

LAWNS FROM TURF

Must I lay turf during the winter?

The only true answer to this would be to say 'no'. If you can keep the turf moist enough for roots to grow then you can lay turf throughout the summer. For ease autumn is probably the best time because nature normally provides just the right conditions for turf to grow. As long as the soil is well prepared and moist the turf will quickly produce roots and become established. Do not lay turf on to frozen ground and avoid droughts if at all possible. Turf which is allowed to dry out shortly after laying will shrink and quickly die off.

You must measure your lawn for turf just as you did for seed. Then you will normally buy turf by the square metre or square yard. Because not every piece will be used, you must buy slightly more than you think you need. You will reject some of the turf because of weeds or bare patches; other pieces will have to be cut or trimmed to size and so there will always be wastage. In normal circumstances you should buy 5–10 per cent more than you need.

Do I need special tools?

You should work on boards when turfing. Obtain enough to cover the width of the lawn. You will probably need a barrow to move the turf from the point of delivery to the lawn.

A garden line is needed to mark straight edges, and a half-moon edging iron – the type used to trim lawn edges – will also be needed. If you don't have one you might find that an old knife or sharp spade will serve just as well.

Method of working

When the soil is fully prepared you should arrange for the turf to be delivered. Have it stacked as close as possible to the lawn to reduce your work load. If the turf can be used within a day or so you can leave it in heaps. If the turf has to be left for three or more days it should be spread out so that every piece is open to the light; failure to do this will result in the turf yellowing and perhaps beginning to rot.

Your turf can arrive in any one of half a dozen different sizes. Some turf comes in pieces 1 m²; other suppliers sell their turf as pieces 90×30 cm², 60×30 cm² and for very fine work 30×30 cm². Long pieces will be rolled up, grass inwards; smaller pieces should be stacked grass to grass and soil to soil.

On your prepared site mark the lawn edge. For straight edges use the garden line, but for curves you might like to use a hose-pipe laid on the ground to the right curve.

Lay turf to form at least two of the edges. (Many people prefer to lay all four sides at the start, but two is often easier.) Try always to use whole pieces of turf to form the lawn edges as these are the strongest pieces and give the best-quality edge. Once two strong sides are formed you should work across the plot and fill in the gap between the two sides. Lay the turf in the same pattern as bricks are laid in house walls – stretcher bond (see Figure 6.4).

a Where turves overlap.

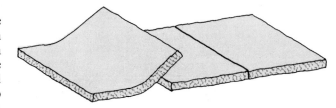

b Using a knife or edging iron cut through the lower piece.

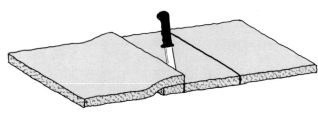

c Remove the lower piece and the upper turf should fit neatly into the gap.

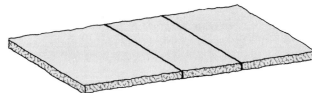

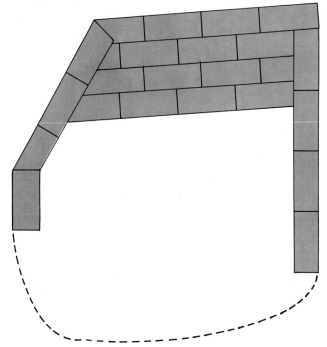

Figure 6.4 Lay turves around the edge of the lawn first of all. Then fill in the centre, laying the turf so that the edges in adjacent rows are staggered like brickwork.

Figure 6.5 Trimming pieces of turf.

Push the pieces of turf as close together as possible; there should be no gaps. Where two pieces of turf meet there will always be some overlap. Fit the pieces by laying one piece over the other and trimming the lower piece to size (as shown in Figure 6.5a and b). Remove the piece of turf which you have just trimmed off and the top piece will then fit neatly into place (Figure 6.5c).

As you lay one strip of turf you should work from a board which has been laid on the previous strip of turf (Figure 6.6). As you complete the second row of turf move the board forward and begin to lay a third row. Each time a row is laid the board will firm the turf into place. Your weight on the board should be all the firming that is necessary; rolling should not be necessary.

Continue to lay the turf strip by strip, continually working from the boards, until the entire lawn is covered.

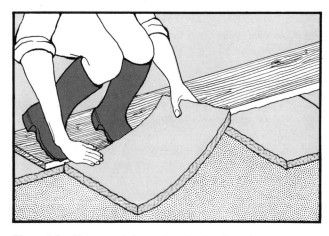

Figure 6.6 Always work from a board to lay the turf.

Figure 6.7 Add soil to fill any hollows below the turf.

Figure 6.8 Use a stiff broom to brush sandy topdressing into the crevices between turves.

Turf which has been lifted by machine should be constant in its dimensions, and if your soil was levelled correctly you should therefore be able to lay the turf quickly. If slight hollows or bumps exist you must add or remove soil to make the finished turf surface smooth (see Figure 6.7). If any turf is too thick it can be trimmed using a turf gauge box as in the diagram in Figure 3.11 on page 22. The turf is laid in the box, grass side down, and a knife is then used to trim off any soil which appears above the edges of the box.

Remove any weeds which may be present. Remember that you should reject the really weedy pieces of turf as they are delivered.

For greater stability on banks, lay the turf across the slope, or you can hold the turf in place by driving pegs through it.

Aftercare

After laying all the turf scatter dry, sandy soil over it. This can then be brushed into any cracks which might exist and will stimulate rooting and will encourage quick growth (see Figure 6.8). In good conditions the joins between the turf should disappear in a few weeks. Keep off the turf completely for the next few weeks and then maintain the lawn as described earlier for seeded areas.

Remember to apply pre-seed fertiliser to ground which is to be sown or turfed as both will benefit from it. If you buy good quality turf such as Rolawn you will find that it has been treated with slow-release fertiliser to encourage quick growth. As the fertiliser wears off and is used up, the grass will become paler and should be stimulated by nitrogen feeding unless winter has arrived. As a general rule do not apply extra nitrogen after the end of August.

CHECKPOINT

Try to answer the following questions without referring to the chapter.
1 What is the best time to sow grass seed?
2 How long should perennial rye grass seed take to germinate in good conditions?
3 How much seed should be sown per square metre to make a fine lawn?
4 Why should young grass plants not be cut very short?
5 In what circumstances should you not lay turf?

Now look back to check your answers.

Before reading this chapter ask yourself the following questions and then bear the answers in mind.

How much time do I have available to care for the lawn?

Do I want a very fine lawn or do I want something which is green and suitable for the children to play or relax on?

THE YEAR AT A GLANCE

I must stress the need for work where lawns are concerned. Anyone can grow grass but it takes skill and interest to produce a lawn. There is always something to do on a lawn unless the weather is so bad that you can't see the grass. A seasonal guide to lawn care is shown in Table 7.1, and Table 7.2 provides an outline guide to which jobs should be done on the lawn and when they should be done. Remember that each table is only a guide. It may not suit your lawn exactly. You must try the programme and then modify it according to the weather or your particular site.

Each month has suggestions for work which might be done, although they may not all be relevant to your lawn. Note that grass should be cut for as long as it continues to grow and that it grows even in the winter during mild spells. Always be prepared to cut the grass, very lightly, if it becomes untidy during the winter months. The grass doesn't stop growing just because the books tell you that the lawn-mower can be put away!

MOWING

The obvious reason why you cut your lawn is because the grass gets too long. But consider these other reasons. Are they true in your case?

a To produce a dense and compact sward.
b To produce an attractively striped piece of smooth turf.
c To produce a surface suitable for childrens' or adults' games.

If the reasons for cutting your grass hinge on it being long or to make it suitable for the children to play on, you are probably interested in producing and also maintaining a hard-wearing lawn. If reasons *a* or *b* are nearer the truth in your case then you must be interested in maintaining the finest of all lawns. Understanding your objectives is vital if the maintenance of your lawn is not to take up an unneccessary amount of your time.

Mowing is perhaps the most important of all the maintenance operations on lawns because it is the one which everyone has to do; it cannot be neglected. If mowing is ignored the lawn develops into a wilderness within a short time. The same cannot be said for fertilising or watering or raking because many lawns are not given any of these; they are simply cut.

Frequency of mowing

How often do you mow your lawn? What does the lawn look like after it has been cut – better or worse than before it was cut? How do you decide when the grass needs to be cut?

I hope that you were perfectly honest with yourself when answering these questions because the answers are vital if you are to improve your lawn. You will find my answers to each of the above questions as you read on through this section of the chapter.

The type of grass which is present in your lawn will, to a large extent, depend on how often you mow as well as how you mow. It may come as a surprise but your system of mowing can actually kill off some grasses yet stimulate others. Irregular mowing with the mower blades set high will undoubtedly encourage the coarser grasses at the expense of the finer types.

You may wish that there was something that you could put on your grass to stop it growing so fast. Some chemicals are available but as yet they have a limited value. Growth retardants such as maleic hydrazide are sometimes used to slow down the rate of growth of grass but they are not yet suitable for fine lawns and there is a risk of scorch and discolouration. It's often better to choose slow-growing grasses than to rely upon chemical retardants. There is a future for the chemicals but some improvements must be made first.

Mowing should be carried out regularly unless the grass is covering a rough area which is out of sight. Domestic lawns should be regularly cared for as this is the only way to guarantee good results.

Table 7.1 **A seasonal guide to lawn care**

For the equipment to carry out each operation shown in this table you should refer back to Chapter 3 where the selection and use of equipment are discussed in detail. This table shows a rather simplified version of a maintenance programme for a lawn. It also provides a good starting point for those who wish to know more about the needs of a lawn. In this chapter I will describe each of the tasks in turn. The programme should be modified to take climate, soil and other factors into account and to suit your own garden. You must get into the habit of observing your lawn, because by so doing you will build up a wealth of knowledge about how effective certain types of maintenance have been.

	Mow	Feed	Water	Scarify	Aerate	Topdress	Pest/disease/ weed control
JANUARY							
FEBRUARY							
MARCH	✓	✓					✓
APRIL	✓	✓		✓			✓
MAY	✓	✓					✓
JUNE	✓	✓					✓
JULY	✓	✓					✓
AUGUST	✓	✓					✓
SEPTEMBER	✓						
OCTOBER	✓						
NOVEMBER	✓						
DECEMBER							

Table 7.2 An outline guide to lawn maintenance month by month

JANUARY
Lightly rake to clear debris. Observe for signs of bad drainage. Overhaul mowers and sharpen tools.

FEBRUARY
Watch for pest/disease attack. Complete turfing. Prepare areas for spring sowing. Apply light topdressing.

MARCH
Mowing should begin this month. Scatter worm-casts. Rake to clear debris. Roll lightly to settle frost lift. Apply wormkillers. Give a light spring feed.

APRIL
Mow more often, remove weeds and resow bare patches. Apply a spring feed if not given in March.

MAY
Reduce the height of cut — in stages. Apply weed-killers in suitable weather. Irrigate, if necessary.

JUNE
Irrigate and feed, as necessary. Lightly rake before mowing. Spike compacted areas.

JULY
Mow regularly. Spike, scarify, feed and irrigate as weather conditions demand.

AUGUST
Give the final nitrogen feed, followed by the final weedkiller. Resow bare patches.

SEPTEMBER
Raise the height of cut. Carry out renovation work — mow, scarify, aerate, resow/turf, topdress.

OCTOBER
Mow as growth demands. Brush very fine lawns whenever possible to remove dew and reduce disease problems. Complete renovation work.

NOVEMBER
Mow as necessary but keep off frosted lawns. Clear fallen leaves. Returf worn areas.

DECEMBER
Continue to brush whenever possible. Clear fallen leaves and complete turfing.

What happens if I mow infrequently?

If grass is allowed to grow long and is then cut very short in one brief encounter with the lawn-mower the result will be very weak grass. Removing too much foliage at one cut results in the roots as well as the shoots being affected. You can actually stunt the roots by such careless mowing. Pale, thin lawns are often the result of this type of treatment. How many lawns have you seen where this is the approach of the gardener?

If the grass has become longer than it should – because of bad weather or because you've been on holiday – do not cut it back to the normal height immediately.

How should I cut the grass?

The correct answer is a little at a time. Remove one-third of the growth immediately and then let the grass recover for four or five days. Then remove another one-third and again leave the grass to recover. In some years the grass may take a long time to get over the shock of mowing so don't be in a rush to cut it again.

Continue to cut regularly, removing one-third at each mowing.

Figure 7.1 The one-third rule.

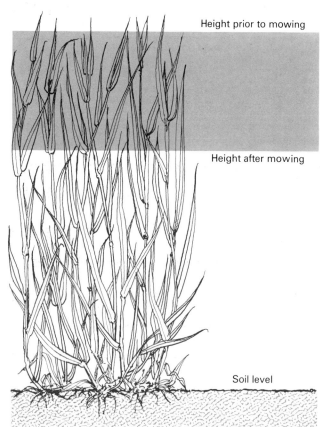

Height prior to mowing

Height after mowing

Soil level

What is so special about one-third?

The one-third rule (see Figure 7.1) has been produced after many years of experience, experiment and exasperation. It is known to be an excellent guide because researchers have shown that to remove more than this amount of leaf at one mowing will actually harm the plant. Remove up to one-third and the plants are not at all harmed.

I do know, however, that there are exceptions to the rule and these exceptions include unusual weather conditions such as drought. The exceptions also include grass which is subjected to shade. This grass should be allowed to grow longer and should be very lightly trimmed.

When to mow

The one-third rule should be your guide. As the weather gets warmer the rate of grass growth will increase and so mowing will be needed more often. As the growth slows down there will be less need for mowing, and in a drought there will be no need to mow at all.

A dry surface makes the task of mowing much easier than a wet surface. It also makes it possible to scatter worm-casts without their smearing over the lawn's surface. If the clippings are being removed (boxed-off) they are easier to collect when they are dry. Wet clippings soon cause the machine to clog.

If the surface of the lawn is very wet or frosted keep off it because you will do more harm than good by treading over the surface.

You should be prepared to mow whenever the grass growth demands it, and remember that this may include winter. The normal mowing season, however, extends from late March to late October, leaving just a little time for mower repairs.

Preparing the lawn for mowing

Surprise, surprise! You thought you simply went out and cut the grass! You can do just that or you can do the job more thoroughly. If you have a very fine lawn you should always do the job more thoroughly.

Grass tends to grow in a matted condition with leaves and stems intertwining. If you cut matted grasses some of the leaves will escape mowing and remain very long. Before mowing it is therefore a good idea to rake the lawn lightly or to brush it with a stiff broom to make the grasses stand upright. They will then be cut cleanly to leave a superior finish to the lawn.

Any worm-casts present on the lawn should be removed by brushing before mowing. If the casts are wet leave them to dry before mowing.

Height of cut

Frequent close mowing will discourage the coarser types of grass and will encourage those grasses which are naturally adapted to survive short cropping by animals. Fescues and bents will therefore survive where perennial rye grass cannot. Cutting the grass very short does have disadvantages though. For example, it can encourage weeds such as pearlwort.

The finest lawns should survive at a height of cut in the 5–6 mm region, but to do so they must be carefully fed and the surface of the lawn must be smooth or it will be scalped by the mower as it moves over bumpy patches.

Coarse, hard-wearing lawns should be maintained at a height of cut in the 18–25 mm region.

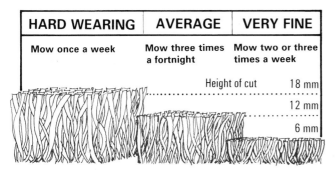

HARD WEARING	AVERAGE	VERY FINE
Mow once a week	Mow three times a fortnight	Mow two or three times a week

Figure 7.2 Rules for mowing.

Winter cutting
Raise the height of cut by a half to strengthen the grass during the winter period, and in spring gradually reduce it again. In droughts the cut should also be raised to reduce water loss.

Deep roots are essential if the grasses are to survive dry spells and harsh winter weather. Allowing the grass to grow a little longer will help to stimulate deep roots. In general the longer the leaves, the deeper the roots.

In poor light, where the grass is shaded, allowing longer leaves to develop will increase the leaf area which can make food and do much to help shaded grass to survive without the need for any special seed mixtures.

Direction of cut
Does the shape and size of your lawn allow you to mow the grass in a number of different directions.? It should.

Try to mow in a different direction every time you cut so that the grass is encouraged to grow evenly rather than in one direction.

For convenience make a turning strip every time you mow. Cut around the outside edge of the lawn first, as shown in Figure 7.3, and then cut the centre of the lawn to leave neat stripes. A machine that doesn't have a rear roller will not give such an attractive finish to the mown lawn as a mower with a roller because the roller makes the stripes.

Collecting the clippings
Professionals refer to this as 'boxing-off' the clippings.

Do you remove the clippings? If so, can you say why? You may be one of the many people who remove clippings because of the mess they would otherwise cause.

Clippings do look untidy and they invariably end up being trampled into the house. There are other reasons for removing the clippings every time you mow:
a it removes weed seeds which would otherwise infest other areas;
b it prevents the formation of unsightly yellow patches;
c it discourages worms;
d it reduces the risk of disease.
In dry spells you can leave the clippings because they can help to reduce the amount of water which is lost from the lawn.

If selective weedkillers have been used don't spread the clippings around trees or shrubs because there may still be enough weedkiller in the clippings to kill other plants.

Leaving the clippings on the lawn will help to recycle some of the essential plant nutrients. It will save you some time but the quality of the lawn may suffer from the smothering effect of the clippings.

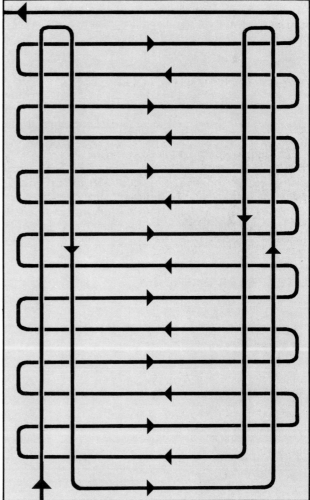

Figure 7.3 Mowing pattern for a striped finish.

Mowing problems

Scalping. This occurs where the surface is irregular and the mower cuts the grass on ridges excessively short, often down to the soil surface. To solve the problem raise the height of cut on the mower or adjust the level of the lawn as described in Chapter 8, 'Lawn problems'.

Ribbing. A series of narrow strips of short grass which alternate with strips of longer grass are referred to as ribbing. This often occurs when the cylinder mower has very few blades on the cylinder or it might occur when long grass is being cut.

Torn grass leaves, those which tend to wither at the tips, are often caused by a poorly set mower, or one with blunt blades which do not cut cleanly. You should have your mower overhauled every year without fail if you want good results and a reliable machine to work with.

FEEDING

Why feed grass? To find out why, try sectioning off part of your lawn and do not give it any fertiliser during the summer. (Use sulphate of ammonia as the summer fertiliser.) Give other sections different amounts of fertiliser – 10 g/m² on one section and 5 g/m² on another – and compare the results at the end of the year.

Grass, like any other plant, requires feeding. Nutrition affects the quality and appearance of the lawn just as mowing does. Failure to feed adequately often results in a sickly-yellow lawn.

The main foods for grasses are nitrogen, phosphate and potassium but some grasses need a lot of food while others need very little. Fescues need very little food and may be killed off if too much phosphate and nitrogen are given. The extra food stimulates other grasses to come into the lawn and replace the fine-leaved types. Table 7.3 shows the main lawn nutrients.

Spring feeding

One complete fertiliser dressing each year should be adequate to meet the needs of most lawns. This can be given either as a spring dressing or as an autumn dressing but rarely is it necessary to give both.

A spring dressing should be applied to a lawn as soon as the grass appears to be growing. If the plants are not active they will not be able to make use of the fertiliser so it's wise to wait.

Proprietary fertilisers are simple to use and save the bother of mixing up a home-made version but they can be rather more expensive. A suitable dressing for 100 m² of lawn could be made up as shown in Table 7.4. (All the materials are easily obtained from gardening shops.) This mixture provides both organic and inorganic forms of nutrient as well as providing quick- and slow-release fertilisers for constant release of nutrients. For simplicity you could miss out the two slow-acting fertilisers (dried blood and bone-meal).

Fertiliser	Amount required (kg)
Sulphate of ammonia	1.5
Dried blood (slow to act)	0.5
Superphosphate	2.0
Bone-meal (slow to act)	0.5
Sulphate of potash	0.5
Calcined sulphate of iron	0.5

Table 7.4 **Mixture of fertilisers for spring feeding**

An alternative spring feed can be provided by using lawn sand. Lawn sand has the advantage of acting as a weedkiller as well as a fertiliser. The mixing and application of lawn sand is discussed under the heading, 'Weed control'.

Table 7.3 **The main lawn nutrients**

Nutrient	Effect on the lawn	When to apply	Source
Nitrogen	Promotes leaf growth and improves colour. A deficiency is seen as leaf yellowing and stunted shoot growth.	Throughout the growing season from spring until mid-August.	Sulphate of ammonia, lawn sand, dried blood and complete feeds.
Phosphate	If a sufficient supply of other nutrients is also available phosphate is able to encourage root growth. A deficiency is seen as a dull blue-green leaf colour and spindly plants with purple leaf margins.	Once a year is adequate in either a spring or an autumn feed.	Bone-meal, superphosphate and complete feeds.
Potassium	Counteracts the effects of excess nitrogen and hardens foliage, thus making it less susceptible to disease attack and cold weather effects. A deficiency is seen as scorch on leaf edges and rolling of the leaf tips.	Once a year as part of a complete feed with phosphate.	Sulphate of potash and complete feeds.

Summer feeding

Nitrogen is required throughout the growing season and it is this nutrient which forms the basis of summer feeds. To maintain colour and vigour during the growing season apply sulphate of ammonia to the lawn at a rate of 17 g/m² at least twice during the summer.

For accuracy in spreading, bulk up the fertiliser with some dry soil or sand, possibly that used for topdressing in the autumn.

Autumn feeding

If a spring fertiliser was given there should be no need for any autumn dressing. Examine the lawn during mid to late August: if the grass cover is thin or weak the lawn will benefit from a light dressing of fertiliser such as that given in spring.

Preparing the lawn for feeding

Prior to the spring or autumn dressings the lawn should be spiked so that the fertiliser can be put close to the roots. Phosphate left on the surface of a lawn is a major cause of surface rooting, so get it down into the soil whenever possible.

If the soil is very hard or dry it should be well irrigated before applying the fertiliser. Irrigation is also beneficial after spreading the fertiliser.

Applying fertilisers

There are two methods to choose from: *a* hand application, and *b* mechanical application.

a Hand application requires great accuracy because some of the materials used can scorch the grass if applied too thickly. Always bulk the fertiliser up with sand or soil to make spreading easier. Mark the lawn into strips and work systematically along each one. For added safety and accuracy divide the fertiliser into two lots, spreading one half lengthways down the lawn and the other half crossways.

b Mechanical spreaders consist of a hopper to hold the fertiliser and a roller or conveyor belt at the base of the hopper to carry fertiliser out. The machine has to be calibrated or set to apply just the right amount of fertiliser. Care is needed to avoid overlap and for safety it is often best to spread the fertiliser in two separate lots, as for hand distribution.

WATERING

In a normal year most well-cared-for lawns will be perfectly able to survive without irrigation for at least eight or nine months. Occasionally, however, there are dry periods and even droughts when the soil reserves of moisture are depleted. In such circumstances irrigation becomes essential.

A well-cared-for lawn – one which is adequately fed,

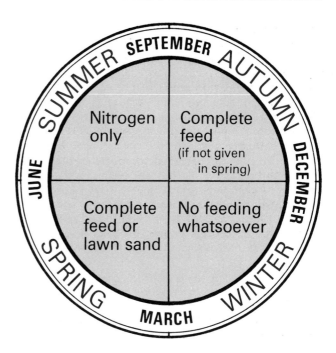

Figure 7.4 Seasonal lawn-feeding programme.

regularly scarified and aerated – will survive long periods of dry weather with very few signs of suffering because the grass has a deep and extensive root system. Such lawns are highly drought-tolerant and hard-wearing.

When to water

A lawn needs to be watered when the symptoms of drought or water stress appear. The colour of the grass becomes dull and has a distinct blue tint. Footprints persist longer than usual because the grass is becoming limp and cannot quickly spring back to the upright position. Eventually the leaves will shrivel and turn brown, the roots will dry up and the plants will die. You should irrigate your lawn when you notice the earliest of these signs.

How much water to apply

The speed with which the symptoms of water stress appear will depend upon soil type. Sandy soils have lower water reserves than clay soils and consequently show stress signs earlier. In hot dry weather a lawn may lose 25 mm of water in 7–10 days. To replace that approximately 27 litres of water would be required for every 1 m² of lawn. You can't measure the water easily but you can still water adequately.

The aim of watering is completely to restore the soil to its full water-holding capacity rather than trying to wet the surface layer only. Frequent light applications of water in dry weather serve only to encourage very shallow rooting and weeds such as moss, with the result that a severe drought soon kills the lawn completely.

As a general rule therefore always water adequately to wet the soil to a depth of at least 15 cm and then allow it to dry out slowly to about 10 cm before watering again. Test your watering by digging a test hole before and after you water. Notice how deep the water penetrates if the sprinkler is turned on for half an hour, one hour and two hours. This is far more accurate than guessing how much to apply.

When very dry, soils accept water only very slowly. If a lot of water is added in a short period of time much of it will remain as pools and will eventually run off to the side. The most efficient method is to sprinkle the surface lightly with water, allow that water to soak in and wet the surface layer, and then add the main supply which should enter the soil much more quickly.

Equipment

Irrigation equipment extends from the very laborious watering can up to the very sophisticated automatic systems. Each system serves a purpose and which is best for any particular lawn depends on the size of the lawn as well as finances available. Methods of watering are covered more fully in Chapter 3, 'Tools and equipment'.

SCARIFYING AND RAKING

Scarification is the vigorous use of a rake or similar piece of equipment to remove weeds and thatch from the surface of the lawn where they form a virtually impenetrable barrier to air, water and fertilisers.

Thatch

Is your lawn soft and spongy? If so it probably has a layer of thatch, a peaty material formed by the grass itself and made up of roots, stems and leaves.

Thatch occurs where the rate of production of organic matter in a lawn exceeds the rate of decomposition. A layer of about 6 mm of thatch is a useful mulch layer to reduce water loss and it acts to reduce the effects of wear on the grass plants. In layers greater than 6 mm deep thatch becomes a barrier to water penetration and it encourages shallow, surface rooting. When the thatch dries out the grass soon suffers from drought, and then when heavy rains occur the thatch acts like a sponge to create a soft, waterlogged lawn.

In recent years excessive thatch has been associated with a number of lawn diseases and weeds. Every effort should be made to control its development and restrict it to a manageable level. You can find further discussion of thatch in Chapter 8, 'Lawn problems'.

Method of scarifying

Press the scarifying rake well into the lawn and then pull it vigorously along the surface of the lawn to pull up as much thatch as possible.

When to scarify

In spring, just as the grass begins to grow, carry out a light scarification to remove any debris left over from winter. After a full growing season there will be much more debris to remove and consequently autumn scarification is much more severe. If mosskillers have been used the dead moss should be pulled out in much the same way as thatch.

Raking is much less severe than scarifying but is basically the same type of operation. A lawn is often subject to an accumulation of leaves from nearby trees. Light raking will remove all such debris and will also lift creeping grasses and weeds so that they can be cut with the mower.

As with scarifying, raking is normally a spring and autumn operation but there is much to be gained from a light raking during the growing season too.

AERATING

Aeration is perhaps the most underused and undervalued operation in lawn care.

Why aerate?

During the year a lawn will be trampled and it will have various machines and pieces of equipment pulled across it. If the soil is wet the weight of the mower or its operator will be enough to begin to compact the soil. Aeration seeks to remedy the situation and to prevent weeds, associated with compacted soil, from becoming established.

If your lawn is hard and holds water for a long time it is probably suffering from excessive compaction. A compact soil is one in which the pore spaces, between soil particles, have been compressed until they have become very small. Small pores tend to hold on to water more strongly than large pores and as compaction increases the soil becomes more and more prone to wetness. A wet soil is readily invaded by weeds and diseases, thatch will develop readily and grass growth will be weak.

During the summer months the soil will not allow water to enter easily and the roots growing in that soil can very soon be faced with drought conditions. The task of aerating the soil is carried out to break through the hard, compacted layers so that water may once again drain freely through the soil and so that oxygen can enter and reach the pores which are close to the growing roots.

When to aerate

How often is your lawn trampled? Daily, weekly, or hourly? The more heavily a lawn is used, the more often it should be aerated. In spring and summer spike compacted areas with a garden fork and repeat the operation approximately every month. The soil should be just moist enough to allow the tines to enter easily. If too wet the soil is smeared as the tines enter and drainage could actually be unaffected even by intensive spiking. In autumn deep aeration is often required. To be wholly effective the tines must break through compact soil to link with well-structured layers.

Aeration is normally a vital part of autumn renovation work. It follows mowing and scarifying in a logical sequence of operations. Once core holes are created top-dressing can be incorporated more easily.

During winter months aeration may still be useful for getting rid of surface water.

Types of tine

There are three main types of tine for use in lawns (see Figure 7.5). The slit or knife-shaped tine is useful for stimulating root growth and for breaking through layers of thatch to hasten its decomposition. The solid tine is of much more value on hard, compacted ground where penetration is difficult.

The third type, the hollow tine, actually removes a core of soil and allows topdressing to be worked into the soil. Hollow tine aeration is especially useful where poor drainage has been a problem as sandy materials can be used to fill the core holes to ensure the rapid removal of surface water.

Each type of tine is normally fitted on to a hand fork. The work is laborious but essential if healthy growth is to be maintained.

ROLLING

The rule to follow regarding rolling is when in doubt, don't. Rolling must surely be one of the most overused and abused operations ever to be carried out on a lawn. It is occasionally very necessary but the emphasis must be placed on the word occasionally.

Rolling, though preferably trampling, a light soil prior to sowing or laying turf is necessary to avoid uneven settlement at a future date. Rolling is also of value in settling frost lift after a cold winter but should never be used to improve the level of an established lawn. Too much rolling or rolling when the soil is wet will only cause compaction and the problems which are associated with it. Poor penetration of water and very shallow root growth are often caused by unnecessary use of the roller.

Traffic over the lawn during the year is usually enough to firm the ground, so much so in fact that any use of the roller may cause serious problems. If in any doubt the roller should not be used at all.

TOPDRESSING

To avoid confusion a definition is required. Where lawns are concerned the term topdressing refers to the application of bulky materials such as soil, sand and peat rather than to the application of concentrated fertiliser-type materials.

Why topdress?

Rolling has already been rejected as the way to improve levels in a lawn because it causes so much unnecessary compaction of the soil. Topdressing is the most effective method of improving minor surface irregularities in the lawn. Hollows can be progressively filled by the addition of small amounts of topdressing materials. Eventually the hollow will be filled and brought up to the same level as the surrounding area.

If poor drainage has been a problem topdressing can help to improve the ease with which water moves through the soil. Permeable materials such as sand should be added to the lawn after thorough aeration has been carried out. The sandy material should be well worked into the lawn so that it fills the tine holes left by aeration work. In a similar way, if the lawn has been too easily affected by drought the ability of the soil to hold water can be improved by adding peat in the topdressing mixture. This operation is shown in Figure 7.6.

In summary we can say that a lawn should be topdressed in order to

a improve the levels of the surface;
b change the nature of the soil and make it hold water;
c improve the drainage of the soil;
d improve the fertility of the soil by adding good quality soil to the lawn;
e help new grass plants to tiller and thicken up the sward.

Topdressing really is a valuable operation, but if your lawn is already perfect don't bother to topdress it. Lawns which are not yet perfect may need topdressing.

You must be observant if you are to satisfy the needs

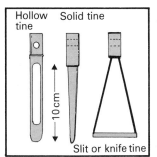

Figure 7.5
Tines for lawn aeration.

Figure 7.6 Aeration enables topdressing to be worked into the core holes to improve the soil. Improved aeration stimulates root growth.

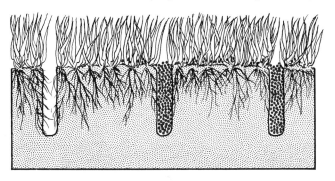

of your lawn. When you identify a problem the following section on materials should help you solve it.

Materials

There are three main ingredients in most topdressing mixtures – loam, sand and peat. A proprietary mixture could be used for convenience but most gardeners seem to prefer to make their own; some even like to use old potting compost.

Loam is ideal as a material to raise levels and it also improves the general fertility of the lawn. The material should be sieved so that all large stones and debris such as plant roots are removed. A 6 mm garden sieve is adequate. A loam stack, made from stacked turves, is the traditional way to produce loam for compost and topdressing but this is not always possible because of scarcity of basic materials.

Sand is the best material to use when drainage has been a problem. Use sand which is salt-free, lime-free and which has been washed to remove silt and clay deposits. The sand should not be coarse in terms of its particle size because such sand fills up easily with silt and thus does little to improve drainage over the long term. A medium to fine sand is ideal, that is one with its particles lying in the size range 0.1–0.5 mm and which might be sold as a soft washed sand.

Organic matter is normally supplied as peat, ideally a sedge peat. Garden compost could also be used just as long as it is well rotted.

Fertilisers are sometimes added to topdressings but this is not always as good an idea as it might seem. Topdressing is added to a lawn and spread carefully to ensure that more is left in the hollows than anywhere else. If fertiliser is added the hollows receive more fertiliser and hence subsequent growth is uneven. If the lawn is already level then it can be assumed to be safe to mix fertiliser with the bulky topdressing.

The correct mixture. There cannot be one mixture to suit every lawn because the climate varies, and soils and lawn maintenance vary. As a general guide, though, the quantities given in Table 7.5 might be useful.

Spreading the topdressing

Topdressing is normally applied as part of the autumn maintenance programme but some lawns may benefit from frequent topdressings throughout the year to raise hollows.

Before you begin, thoroughly prepare your lawn by close mowing, scarifying and spiking. Then mark the lawn off into 1 m wide strips using garden lines.

Allocate topdressing to each strip in turn at a rate of about 1–2 kg/m², although there is no hard-and-fast rule. Spread the topdressing along each strip using a shovel (see Figure 7.7). You'll find that a gentle sweeping action with the shovel kept close to the ground will ensure fairly even distribution.

To improve heavy soils	Materials	To improve light soils
	PEAT	
	LOAM	
	SAND	

Table 7.5 **Topdressing mixtures**

Figure 7.7 Spread topdressing by shovel after first marking the lawn off into strips.

When you've topdressed each strip you can remove the lines to allow the next stage to take place. Work the dressing into the grass with a stiff brush, besom or lute. The back of a rake would also be suitable. When you've completely worked the dressing into the lawn there should be no risk of it smothering the grass or encouraging disease.

You can now leave the lawn for several days during which time the grass will grow through the dressing.

PEST, DISEASE AND WEED CONTROL

These three topics will be dealt with together initially but specific problems will be discussed in more detail in Chapter 8, 'Lawn problems'.

The key to pest, disease and weed control is prevention. The most effective method of reducing these major problems is to ensure a healthy and vigorous lawn by attending to all other items of maintenance. A strong, dense lawn can compete with weeds and is less easily damaged or infected by disease. A poor lawn – one which is weakened by over- or under-feeding – is likely to suffer from annual disease attacks and weed invasion.

Weed prevention
The most troublesome weeds are those which survive for several years, the perennials, and which are low-growing thus avoiding the blades of the mower. Annual and biennial weeds normally die off as mowing continues and so are of little concern for the moment.

A clean seed-bed is the ideal starting place to ensure a clean lawn. From there it is necessary to be vigilant to kill off weeds as they appear rather than allowing a high population to build up.

Box-off grass clippings whenever possible to avoid spreading seeds to other parts of the lawn. When weeds are allowed to establish in the lawn they tend to smother grasses and create bare soil. If weeds are killed off the bare soil must be recolonised fairly quickly by new grass seedlings otherwise weeds will reappear.

Weedkillers and specific weed problems are dealt with fully in Chapter 8, 'Lawn problems'.

Disease prevention
Good maintenance is again the key to avoiding diseases in lawns. Most turf diseases are encountered on lawns which are poorly cared for, though even the most well-looked-after lawn is at some risk of infection.

Fungi cause the most serious lawn diseases. Free water must be available on the leaf surface if the fungal spores are to be able to germinate, grow and infect the grass. Regular maintenance work should aim to create as dry a surface as possible to inhibit the growth and development of fungi.

Good drainage is essential but regular brushing to remove heavy dew formation is also of value. A long cane (switch) or brush will be adequate if dragged lightly across the grass surface in early morning.

Several fungi affect lawns. Their identification and control are dealt with, along with pests of lawns, in Chapter 8, 'Lawn problems'.

THE NEGLECTED LAWN
Renovating a lawn which has been sadly neglected for some years is no easy task. Patience and hard work are essential if the results are to be worth while. This programme assumes a spring start to the work.

Stage 1. Examine the lawn carefully. If most of the area is weedy and there are few signs of healthy grasses then consider starting again by making a new lawn. If such is the case refer to Chapters 2 and 4 which deal with the construction and planning of new lawns.

Stage 2. Assuming that there is sufficient grass alive to make renovation possible, the whole area should be checked for stones, obstructions and other debris. Remove anything which might cause damage to machines or tools.

Stage 3. Cut down the grass and weeds to approximately 5–7 cm. Ideally use a large rotary mower which is specially designed for rough grass areas. If such a machine cannot be hired a scythe will serve the purpose. Rake off all long grass and other vegetation.

Stage 4. Check the lawn for uneven spots, weeds, etc., and make a note of those which cannot be dealt with immediately. The common troubles are dealt with in Chapter 8.

Stage 5. During the summer cut the grass as often as possible with the blades of the mower set to 25–50 mm. As the grass begins to recover the blades of the mower can be progressively lowered until the preferred height of cut is reached.

Stage 6. In the early part of summer give the area a dressing of compound or complete fertiliser as discussed in this chapter. Treat weeds with a selective herbicide.

Stage 7. By late summer or early autumn the lawn should have improved considerably. Any pests or diseases can now be dealt with and uneven spots or bare patches can be dealt with as detailed in Chapter 8.

Stage 8. Once any major problems have been eliminated the lawn can be further improved by following the normal maintenance programme detailed earlier in the chapter.

HOW MUCH TIME IS NEEDED?
How much time can you devote each week to the lawn? Your answer will determine the quality of lawn that you can produce. Table 7.6 gives estimates of the work needed for 100 m² of fine turf in the course of one year.

Table 7.6 **Times spent each year on maintenance (100 m² fine turf)**

Operation	Approximate number of times performed in the year	Total time/ year (hours)
Mowing with a 35 cm mower	30	4
Topdressing	1	1
Brushing	1 (after topdressing)	0.5
Spiking	2	5
Scarifying	2 (vigorously)	2
Trimming edges	10	6
Adding fertiliser	4	4

Operation	Area/distance	Time required
Brushing	100 m²	30 min
Topdressing	100 m²	1 h
Aeration by hand fork (spiking)	100 m²	2.5 h
Scarifying by hand rake	100 m²	1 h
Trimming lawn edges with shears and collecting trimmings	100 m	1.5 h
Laying turf	100 m²	5 h
Marking, cutting and lifting turf	1 m²	10 min
Grass cutting: 35 cm cylinder	100 m²	8 min
90 cm cylinder	100 m²	3 min
45 cm rotary	100 m²	10 min
Digging and weeding a site	100 m²	6 h
Edging a lawn with a half-moon iron	40 m	1 h
Raking to level a bed	100 m²	2 h
Raking a tilth for seed-sowing	100 m²	2.5 h
Rotavating ground with a 45 cm machine	100 m²	35 min
Applying wormkiller by watering can	100 m²	30 min
Adding fertiliser by hand	100 m²	1 h

Table 7.7 **Times required for construction and maintenance operations**

Table 7.7 provides some times required for the operations carried out as part of the construction and maintenance of a domestic lawn.

Without counting the use of weedkillers, you might spend up to 22.5 hours on a small lawn during the course of a year. In a week during the summer you may have to cut twice and feed once. Watering should take up very little time except to set the sprinkler up, but in one busy summer week you may still spend up to three hours just on the lawn.

CHECKPOINT
Try to answer the following questions without referring to the chapter.
1 In what circumstances should a lawn be allowed to grow longer than normal?
2 What are the main nutrients in a lawn-feeding programme?
3 What effect has aeration on a lawn?
4 What problems are associated with frequent, light irrigation of lawns?

Now look back to check your answers.

CHAPTER EIGHT

Lawn problems

Bare patches/Thin, sparse grass/Brown patches
Yellow or pale green areas/Weeds/Uneven surface
Turf diseases/Waterlogging/Spongy surface
Black slime on the surface/Pest damage/Damaged edges
Coarse grasses/Stones/Cracks between new turves
Cracks in the soil/Checkpoint

Invariably something will go wrong with your lawn eventually, and although it may not be serious it can still be worrying. This chapter should help you to identify the cause of the most common problems and show you how to deal with them.

Go out and examine your lawn. Does it look perfect or does it exhibit one or more of the symptoms given in Table 8.1?

Each section in this chapter will provide some information about the cause of a particular problem and suggest some possible remedies. You may then be directed to another part of the book if it is also relevant.

Symptom	Chapter section
Bare patches	I
Thin, sparse grass	II
Brown patches	III
Yellow or pale green areas	IV
Weeds	V
Uneven surface	VI
Fairy rings or other diseases	VII
Waterlogging	VIII
Spongy surface	IX
Black slime on the surface	X
Signs of pest damage	XI
Damaged edges	XII
Coarse grasses	XIII
Stones appearing on the surface	XIV
Cracks between new turves	XV
Cracks in the soil	XVI

Table 8.1 **Chapter sections in which lawn problems are discussed**

Section I: **BARE PATCHES**

Possible causes
a Failure of seed to germinate because of disease.
b Poor seed distribution during sowing.
c Poor seed germination because of poor soil or a compacted, waterlogged soil.
d Poor germination because you used very old seed.
e Petrol or chemical spillage.

f Double-dosing the area with fertiliser.
g Heavily compacted soil.
h Too much wear and tear.
i Scalping by the lawn-mower.

Suggested action
a Treat the seed with a fungicidal seed dressing prior to sowing and then resow the patch.
b Resow the patch.
c Aerate the soil thoroughly prior to sowing.
d Buy fresh seed and discard old seed.
e Remove the affected soil and turf and replace it as described below and illustrated in Figure 8.1.
f Water the area well on several occasions and then resow the patch or wait for natural regrowth.
g Aerate the soil prior to resowing.
h Reduce the use of the lawn until the grass has had time to recover.
i Adjust the levels of the lawn (See Section VI).

Bare patches can be resown or returfed at any time of the year as long as the soil is warm and you can keep it moist.

Cut out and remove the affected soil or turf and lightly loosen the rest of the soil with a fork. Add a little pre-seed fertiliser to the soil and then firm it by treading. Cut fresh turf to fit the gap, position it carefully, firm the turf and then topdress with sandy soil to fill the cracks (see Figure 8.1).

If seed is to be used you must add fresh soil to fill the hole, firm the soil and then scatter seed over the soil. Cover the seed lightly with soil and then cover the area with cotton or a net to keep the birds away.

Section II: **THIN, SPARSE GRASS**
Thin, sparse grass on a new lawn may be caused by poor seed-bed preparation. Other factors – shade, for example – could also cause the same symptoms.

Severe soil compaction can produce the symptoms but so can poor mowing. If compaction is a problem intensive aeration will overcome it. Refer to Chapter 7, 'Looking after established lawns', for further discussion.

Lack of adequate feeding can result in a thin sward. Applying a balanced fertiliser as described in Chapter 7 will solve the problem.

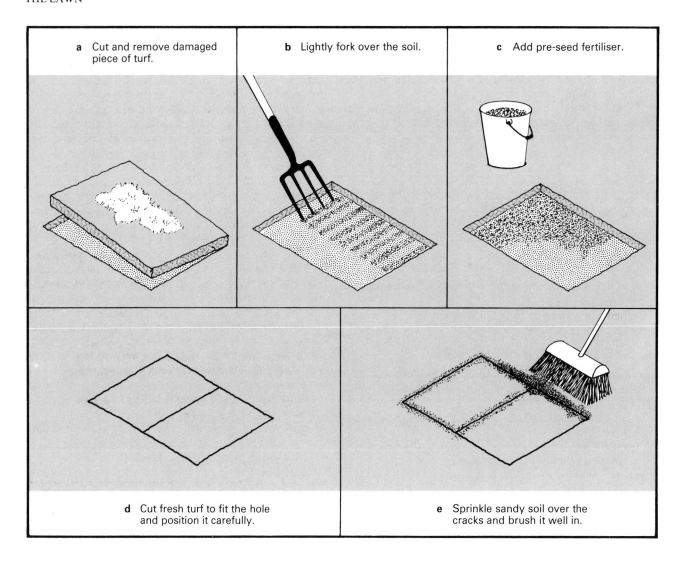

a Cut and remove damaged piece of turf.

b Lightly fork over the soil.

c Add pre-seed fertiliser.

d Cut fresh turf to fit the hole and position it carefully.

e Sprinkle sandy soil over the cracks and brush it well in.

Figure 8.1   Repairing a bare patch in the lawn.

Section III: **BROWN PATCHES**

There are many possible causes of such patches on lawns.

Straw-coloured patches which have either a pink or a whitish tinge or mould covering them will probably be caused by red thread or fusarium diseases. If only fescue grasses are affected you should check the notes covering dollar spot. All of the most common diseases are covered in Section VII.

If the grass in the brown patch has had its roots eaten then you will probably find that the cause is leatherjacket and this pest is dealt with in Section XI.

What else could cause brown patches?

Before you assume that diseases have caused the patches on your lawn you should realise that other causes exist. Bitch urine is commonly to blame for damage to lawns.

There is often a dark green ring around the patch. Look out for this sign.

Rubble just under the soil surface, shallow soil over subsoil, oil spillage and double-dosing with fertiliser can cause brown patches too. Replace the soil if possible and resow or returf the affected patch (as explained in Section I).

Section IV: **YELLOW OR PALE GREEN AREAS**

Poor colour in a lawn can indicate one of several problems. Usually the problem is the lack of sufficient fertiliser but not always.

If you haven't fed the grass for a long time then a light feed, as described in Chapter 7, should help. It is often worth while just digging a shallow hole in the lawn to examine the roots and the soil.

Are the roots very shallow? Shallow roots can indicate

compaction and waterlogging. You will find both discussed in Section VIII of this chapter.

Have the roots been eaten? Roots which just suddenly stop near the soil surface may have been eaten by pests. Refer to Section XI.

Does the soil have a thatch or a spongy layer present at the surface? Thatch is dealt with in Section IX.

Section V: **WEEDS**
The commonest definition of a weed is a plant growing in the wrong place.

Methods of control
Where control does not involve the use of chemicals or weedkillers the methods are referred to as cultural controls. Such methods should always be the first consideration in keeping weeds down.

When a new lawn is being constructed all weed growth should be carefully removed or killed using a suitable herbicide. Fallow the ground for as long as possible so that you have the opportunity to kill the weeds as they germinate. There is always a good supply of weed seed in the soil ready to germinate when the conditions are favourable.

If you are buying turf always check the supply for weed infestation.

Coarse grasses may often be killed by regular scarifying or slashing with a knife. Creeping weeds can often be controlled in established lawns by regular raking and scarifying.

If your ground preparation and maintenance are carried out properly then you will minimise the weed problem for the future. Too much or too little water or fertiliser may cause weeds to encroach into your lawn.

When all else fails chemical controls must be your choice. The lawn-mower will kill off many weeds – mainly the annual and biennial types – but others will persist and must be dealt with properly.

Why are weeds harmful?
Simply put, they compete with the grasses for food, water, light and air. They are often very aggressive and can weaken the grass in which they are growing. Eventually the weeds become dominant and the grass dies off completely. A few flowers in the lawn might look nice initially but inevitably there will be more and more of them.

What types of weedkiller are there for use on lawns?
When a new seed-bed is being cleared you may use a *contact herbicide*, one which kills the parts of the plants which it touches. Residual types last in the soil for some time and will kill anything attempting to grow there. These types are not of much value to us. *Translocated herbicides* are those which move within the plant so that they reach

and kill even the roots; contact herbicides cannot do this. *Total weedkillers* kill everything; *selective weedkillers* kill only some plants. On a lawn it is quite common to use selective weedkillers which kill the broad-leaved weeds but selectively leave the grasses unharmed.

When using selective weedkillers always read and follow the instructions on the container. Never add a little more for luck as it always proves to be unlucky and could result in the grass and everything else dying.

The quickest way to get results is to use the chemicals when all the plants, both grass and weeds, are growing vigorously. To help matters along apply a nitrogen fertiliser about 7–10 days before applying weedkiller to the lawn. This will boost the grass and hasten the death of the weeds.

If the lawn is heavily infested with weeds you should apply the chemical by watering can or sprayer. Mark the lawn into strips for ease and accuracy and then spray each strip in turn.

If there are few weeds you might find it useful to buy an aerosol can of weedkiller so that you can spot-treat the weeds. You can always paint individual weeds with weedkiller on a brush, but be careful not to spill the chemical and always follow the golden rules of weedkilling.

a Always read and follow the instructions carefully.
b Always use the recommended dose.

Table 8.2 **Weed control table**

Common name of weed	Botanical name	Chemical to use
Creeping buttercup	*Ranunculus repens*	MCPA, 2,4-D
Common chickweed	*Stellaria media*	2,4-D + Mecoprop
Pearlwort	*Sagina procumbens*	Mecoprop
Parsley piert**	*Aphanes arvensis*	Mecoprop
White clover**	*Trifolium repens*	Mecoprop
Plantain	*Plantago* spp.	MCPA, 2,4-D, Mecoprop
Daisy**	*Bellis perennis*	2,4-D
Chamomile	*Chamaemelum nobile*	2,4-D, MCPA
Yarrow**	*Achillea millefolium*	Mecoprop + 2,4-D
Cat's ear**	*Hypochoeris radicata*	2,4-D, Mecoprop
Dandelion	*Taraxacum officinale*	2,4-D, MCPA

*Weeds marked ** may require a repeat spray for full control.*

c Never spray during times of poor weather, drought, wind or high temperatures.
d Do not mow just before spraying.
e Store weedkillers safely, in labelled bottles, well away from children.
f Keep chemicals away from pets.
g Wear rubber gloves to handle the chemical.
h After spraying, thoroughly wash yourself and your sprayer or can.
i Use a watering can with a dribble bar fitted if there is a risk of spray drifting on to nearby plants.
j Never spray seedling grass.
k Don't use the clippings from recently sprayed lawns as a mulch.

Some weeds tend to be difficult to kill and may need a repeat spray to be killed. Remember not to be too impatient – some herbicides take several weeks to be fully effective.

Table 8.2 provides a guide to the main lawn weeds and how they might be controlled. Following this, Table 8.3 describes some of the herbicides which might be of special interest to you.

Special weed problems
Moss in your lawn can be one of three types, the upright type (*Polytrichum*) though this is rarely a problem, the tufted type (*Ceratodon*) which is a particular problem on acid soils, or the trailing type (*Hypnum* and *Eurynchium*) which is commonly present in soft, spongy turf. Mosses commonly appear in spring and autumn.

Why does moss come into a lawn?
The presence of moss is normally a sign that something is wrong with the conditions under which the grasses are growing. If the grasses are weak their place will be taken by mosses and so you rarely see moss on a healthy lawn.

What favours moss invasion?
The conditions which favour moss are those which discourage grasses. These are compaction, poor drainage, very high or low pH, excessive thatch, and shade.

If moss appears in your lawn you should determine what has encouraged it to appear rather than spending your money on chemical mosskillers. If you can find out what is wrong with your lawn you should endeavour to correct the problem first and then kill the moss. Using chemicals without correcting the problem may be quite effective in the short term but the moss will always come back.

Chemical controls include the use of lawn sand which you can make by mixing the following:

20 parts of sand
3 parts of sulphate of ammonia ⎫ all parts by weight
1 part of sulphate of iron ⎭

This mixture should then be spread on to the lawn at a rate of 140 g/m². Choose a day when there is a heavy dew because this helps the chemicals to stick to the leaves of the weeds. Keep off the grass for the next two days otherwise you will leave black footprints everywhere. If there is no rain during those two days you should water the lawn sand into the lawn. After this the moss and other weeds will turn black and must be raked out. Lawn sand contains fertiliser and will stimulate the grass to recover quickly and fill in the gaps left by the weeds.

If you have no wish to make up lawn sand there are proprietary chemicals on the market which are very good. Buy, for preference, a pack which contains a mosskiller based on dichlorophen.

Table 8.3 **Useful herbicides for lawns**

Herbicide	Suitability and action
Ioxynil, Bromoxynil	Both are suitable for use on seedling lawns once plants have two true leaves or more. Before this stage do not use any herbicide. Contact action.
Paraquat	Contact herbicide useful for killing emerging weeds. Is not selective and so cannot be used on established lawns. Suitable for ground clearance prior to sowing.
Mecoprop, 2,4-D, MCPA	Selective herbicides suitable for use on established lawns to kill broad-leaved weeds. There is usually a residual effect in the soil for several weeks after spraying.
Glyphosate	Suitable for controlling annual and perennial grasses and broad-leaved weeds. Very useful for ground clearance during the early stages of lawn construction. Glyphosate rapidly becomes inactive in the soil and sowing and planting work can follow just a few days after spraying.

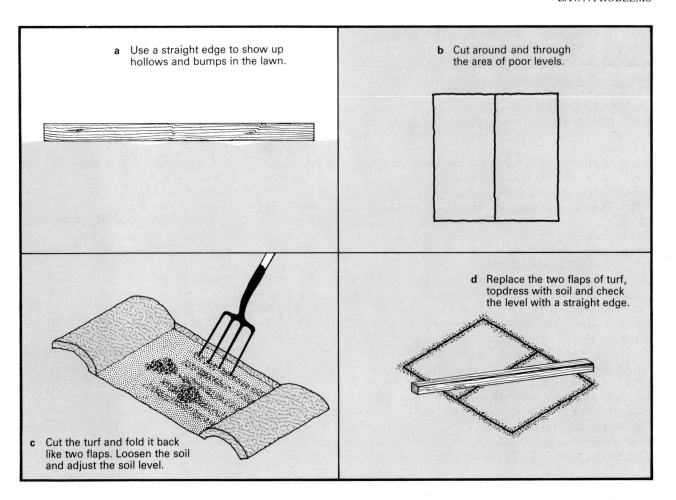

a Use a straight edge to show up hollows and bumps in the lawn.

b Cut around and through the area of poor levels.

c Cut the turf and fold it back like two flaps. Loosen the soil and adjust the soil level.

d Replace the two flaps of turf, topdress with soil and check the level with a straight edge.

Figure 8.2 Levelling the lawn surface.

Section VI: **UNEVEN SURFACE**

Bumps and hollows develop as the lawn settles and compacts. All new lawns will settle to some extent but good seed-bed preparation will reduce it to a minimum. Organic matter in the soil eventually rots and as it does the ground will settle and sink lower. This is one of the reasons for removing old roots from the lawn site during construction.

The action you should take (summarised in Figure 8.2) is as follows:

a Check the extent of the unevenness by placing a straight edge over the uneven area.
b Mark off the area to be improved, cut the turf and fold it back like two flaps.
c Loosen the soil with a fork, and then add or remove soil according to whether the area is high or low.
d Firm the soil by treading, rake it level and replace the turf.
e Topdress to fill in the cracks and then check that the turf is level by using the straight edge again.

Section VII: **TURF DISEASES**

Fungi

The most common turf diseases are caused by fungi. The early symptoms of these diseases are very similar and it is easy to confuse one disease with another. If this becomes your problem use a broad-acting fungicide which will kill the disease, whether or not you know its name.

The easiest disease to recognise is the fairy ring, but did you know that there are three different types?

Grade 3. The least significant type only produces a ring of toadstools and there is rarely any need to do anything at all because there is no damage to the grass. Magnesium sulphate or iron sulphate is sometimes used to slow down the growth of the fruiting bodies.

67

Grade 2. This simply consists of a dark green ring of stimulated grass, possibly with fruiting bodies at the edge of it. Apply nitrogen fertiliser to hide the dark ring, i.e. make all the grass dark green.

Grade 1. The most serious ring of all is caused by *Marasmius oreades*, a fungus which, like the others in its group, feeds on organic matter in the soil. As the fungus feeds on the organic matter it releases nitrate into the soil as a waste product. The grass nearby takes up the nitrate as a food and is stimulated to grow quickly and become darker in colour. This is how the green rings are formed.

Marasmius actually forms two rings, as Figure 8.3 shows. The space between the rings is full of dead grass. Death occurs because the soil is full of fungal mycelium which prevents water entering. The grass therefore dies because of drought.

The fungus itself will die. The mycelium rots in the soil and more nitrate is released to be taken up by the grass nearby, hence the second ring of darker grass.

The old remedy was to dig out the infected turf, sterilise the soil and returf the area. Since 1983 professional gardeners have been able to buy Ringmaster, a fungicide based on oxycarboxin, which actually kills fairy rings and is a great step forward. This chemical should be available to amateurs within a few years. Fairy rings normally appear late in the summer or autumn.

Other turf diseases

Many other diseases of turf appear as brown or straw-coloured patches. Examine the patches for the tell-tale features which help to identify each disease:

Is there a white mould on the turf? YES: Fusarium.
Can you see any pink needles growing from the grass leaves? YES: Red thread.
Is only fescue affected? YES: Dollar spot.
Is seedling grass affected? YES: Damping off.

Fusarium patch disease. This normally occurs at the end of the growing season when growth is slowing down. The first signs are small brown, circular patches, about 10 cm in diameter. A pink or white mould may cover the grass and eventually the affected grass will collapse and become slimy.

Overfed turf is especially susceptible to fusarium so do not feed grass with nitrogen late in the year. Fine turf is often affected but the grass is not always killed.

Use chemicals which contain quintozene or benomyl for full control of the disease. A soil with a high pH is also particularly susceptible so avoid using lime as far as possible.

Red thread disease is caused by *Corticium fuciforme*. The first signs are brown patches but pink needles are usually seen growing out of the grass leaves. If the needles break off they can remain dormant until favourable conditions occur. They then germinate and set up a new infection.

Red thread is especially common on light, sandy soils which are deficient in nitrogen. Spread of the disease can be especially quick in mild, damp weather. This disease may be seen at any time from June to December but then normally disappears.

Figure 8.3   Grade 1 fungus.

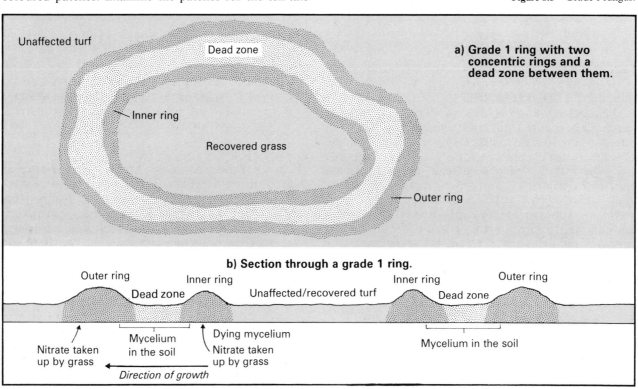

a) **Grade 1 ring with two concentric rings and a dead zone between them.**

Unaffected turf

Dead zone

Inner ring

Recovered grass

Outer ring

b) **Section through a grade 1 ring.**

Outer ring   Inner ring   Inner ring   Outer ring
Dead zone   Unaffected/recovered turf   Dead zone

Nitrate taken up by grass   Mycelium in the soil   Dying mycelium   Nitrate taken up by grass   Mycelium in the soil

Direction of growth

Control is best achieved by a good feeding programme during the year, but chemicals such as benomyl can be used as a last resort.

Dollar spot, as its name suggests, first appears as small brown spots, about 10 mm in diameter, but they increase to 50 or 75 mm in diameter very quickly. This disease appears in summer but usually dies out in winter.

Fescue is especially prone to attack and wet soils, high humidity and weak turf favour spread of the disease.

To control the disease keep the turf adequately fed and the soil well drained. Quintozene and benomyl can be used as a last resort.

Damping off disease can be caused by several fungi which rot the young plants at soil level so that they collapse and die. Sowing seed too thickly is the most common cause but poor drainage contributes to the problem.

Good seed-bed preparation and careful sowing are the best controls but you can buy a seed dressing to discourage the disease.

Section VIII: **WATERLOGGING**
Water standing on the surface of the lawn is often a sign that the soil is excessively compacted and that water cannot escape through it to the subsoil or drains.

If the soil is hard and compact you must carry out intensive aeration using a hollow tine fork if you have one. A garden fork will help but is not as effective. If the waterlogging persists after aeration you should examine the site carefully as described in Chapter 4.

Section IX: **SPONGY SURFACE**
Is the surface of your lawn very soft, and does it hold water for a long time after rain?

Dig a small, shallow hole in the lawn. Is there a layer of peat-like material on the surface of the soil? Does walking across the lawn leave large footprint marks for a long time afterwards?

If your answer to these questions is 'yes' your lawn has a thatch problem.

Thatch is organic matter produced by the plants in your lawn. The lawn is producing this material faster than bacteria and fungi can break it down. The result is a gradual build-up of the layer. More than 6 mm can be harmful because it stops fertiliser and topdressing getting down to the plant roots. Thick thatch layers tend to trap water; the thatch becomes wet but the soil below remains dry. Your grass roots will stay in the thatch rather than grow in dry soil and so your lawn becomes very shallow-rooted. Drought quickly causes the grass to suffer.

Compaction is often a cause of thatch building up but so is poor watering. Giving a lawn a little water very often tends to make it stay wet, an ideal condition for thatch

to build up. Allowing the lawn to dry out between waterings is very useful.

If thatch becomes a problem carry out intensive aeration and scarifying as described in Chapter 7.

Section X: **BLACK SLIME ON THE SURFACE**
Black or black-green slime on the lawn indicates the presence of algae and these suggest a thin turf and a waterlogged soil.

Avoid rolling the soil and carry out intensive aeration.

Lawn sand – see weedkiller notes in Section V – will get rid of the algae but so will copper sulphate: use 34 g in 136 litres of water to treat 100 m².

Section XI: **PEST DAMAGE**
The main turf pests are earthworms and leatherjackets.

Earthworms
Worms are regarded as pests because of their casts, which are unsightly and which are often smeared to leave a patch of bare soil which encourages weed entry. Wet casts produce a slippery surface but the soil may blunt the mower blades making casts a costly problem. Moles may be encouraged to come into the garden in search of the worms and they cause terrible upheaval.

Worms in the soil do a great deal of good but on fine lawns they cannot be acceptable. It is possible to discourage them by:
a collecting the clippings from mowing;
b maintaining a low soil pH;
c avoiding lime;
d using acid fertilisers such as sulphate of ammonia.
Casts are likely to appear in mild, damp, autumn weather. At this time of the year you can carry out chemical control of the worms using carbaryl.

Leatherjackets
Leatherjackets are the larvae of the crane-fly (*Tipula*) or daddy-long-legs. Eggs are laid in the turf during the summer and in the following spring the larvae feed actively on the roots of the grasses. The main feeding periods are from March to May and throughout the winter. Starlings often peck feverishly through lawns in search of the larvae. If starlings like your lawn it probably has a leatherjacket population.

Regular aeration reduces the leatherjacket problem in lawns but where attacks are serious you might try HCH watered into the turf as this gives some control. You might also like to try the old method of covering the turf with a plastic sheet overnight. In the morning you can remove the sheet to find the larvae on the surface. Pick them up and dispose of them.

Other pests

Moles may burrow into your lawn in search of worms. All sorts of remedies are suggested and all have had some success. Traps are effective if positioned properly in the runs. The most efficient method is to call in a professional mole catcher. The problem is that if mole-infested land adjoins your own there can be no long-term solution.

Dogs. Brown patches on lawns caused by bitch urine are fairly common. Water the patch to dilute the effects and be prepared to resow or returf it.

Chafer grubs are curved grubs which feed on grass shoots during spring and summer. Carbaryl is often effective against them.

Section XII: **DAMAGED EDGES**

Repair these by cutting out the affected turf, trimming the edge, turning the turf to make a neat edge to the lawn and filling up the gap with compost and seed, as described in Figure 8.4.

Figure 8.4 Repairing lawn edges.

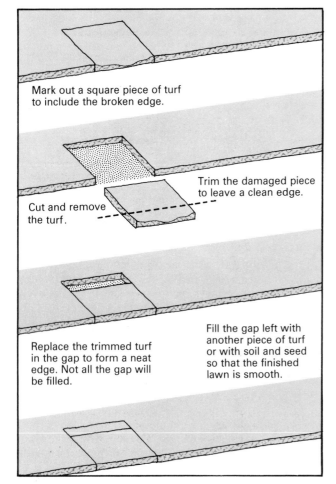

Mark out a square piece of turf to include the broken edge.

Cut and remove the turf.

Trim the damaged piece to leave a clean edge.

Replace the trimmed turf in the gap to form a neat edge. Not all the gap will be filled.

Fill the gap left with another piece of turf or with soil and seed so that the finished lawn is smooth.

Section XIII: **COARSE GRASSES**

Invasion by coarse grasses often signifies that the lawn is being neglected in some way, but excess watering or feeding can encourage such grasses. Raising the height of cut or cutting too infrequently may also contribute to the problem.

Control can be as easy as reversing what you are doing wrong, but unfortunately it rarely is. Regularly scarifying and mowing will help to kill off the grasses, and allowing the lawn to become dry in summer between waterings will also help. Try slashing the clumps of grass with a knife if you haven't a scarifier.

Section XIV: **STONES**

Stones on the surface of the lawn will have worked their way up from the lower layers of the soil. A light rolling will press them down out of the way of the mower blade but you should try to collect them as they appear.

Section XV: **CRACKS BETWEEN NEW TURVES**

Such cracks occur when the turf has been allowed to dry out after laying. The problem is more likely to occur if the cracks were not filled with sandy topdressing immediately after laying.

To repair the damage, water the turf and topdress it to fill in the cracks. Keep the turf well watered afterwards until it is fully established.

Section XVI: **CRACKS IN THE SOIL**

This is the risk to which a spring sowing will always be subjected, if the soil dries out.

Water the soil thoroughly and then topdress the lawn to fill in the cracks. If the grass is damaged seriously add a little more seed to thicken up the lawn. Keep the soil well watered during dry weather.

CHECKPOINT

Try to answer the following questions without referring to the chapter.

1 What weedkiller can safely be used on young grass?
2 What is a selective weedkiller? Can you name one?
3 What are the ideal conditions for spraying weedkiller?
4 How would you recognise red thread disease?
5 What causes the dark green rings of grass in a fairy ring?
6 Why are earthworms regarded as pests?
7 Why are leatherjackets pests?

Now look back to check your answers.

Index